Basic Concepts of Graph Algorithms

Combinatorics for Computer Science (Units 6 and 7)

S. Gill Williamson

Preface

From 1970 to 1990 I ran a graduate seminar on algebraic and algorithmic combinatorics in the Department of Mathematics, UCSD. From 1972 to 1990 algorithmic combinatorics became the principal topic. The seminar notes from 1970 to 1985 were combined and published as a book, *Combinatorics for Computer Science (CCS)*, published by Computer Science Press. Each of the "units of study" from the seminar became a chapter in this book.

Here, we isolate a combined Unit 6 and Unit 7, corresponding to Chapter 6 and Chapter 7 of *CCS*, and reconstruct the original very helpful unit specific index associated with these two units. Theorems, figures, etc., are numbered sequentially: DEFINITION 6.10 and EXERCISE 6.29 refer to numbered items 10 and 29 of Unit 6 (or Chapter 6 in *CCS*). Unit 6 contains basic material at an introductory level. Unit 7 applies Unit 6 to a more advanced topic (planarity testing).

These notes focus on the visualization of algorithms through the use of graphical and pictorial methods. This approach is both fun and powerful, preparing you to invent your own algorithms for a wide range of problems. For further references and ongoing research, search the Web, particularly Wikipedia and the mathematics arXiv (arXiv.org).

Also available in this series are *Basic Concepts of Linear Order* (Unit 1), *Sorting and Listing* (Unit 2 and Unit 3)), and *Pólya Counting Theory* (Unit 4). Units 6 and 7 are essentially independent of earlier units.

The exercises in this material were designed for student presentation in the seminar. In many cases, these presentations were done after we had gone through the entire unit. A good strategy is to read and understand these exercises and return to the ones that interest you after you have read the unit.

S. Gill Williamson, 2012
http : \www.cse.ucsd.edu\ ∼ gill

Table of Contents

v

Unit 6

Basic Concepts of Graph Algorithms

In this chapter, we study a selection of topics from a vast area of combinatorial theory called "graph theory." One definition of a graph is that a graph G is a pair of sets (V,E) where V is arbitrary and E is a subset of the set of all subsets of E of size 2. Never mind for the moment if this definition seems unclear. Note, however, that E may be taken as the empty set ϕ. Thus, all pairs (V,ϕ), where V is any set whatsoever, are graphs by this standard definition. These sets are in obvious correspondence with all sets V $\Leftrightarrow$ (V,ϕ). Thus, to embark blindly on the task of trying to learn all there is to know about graphs in general (without some application or outside motivation in mind) is to learn all there is to know about arbitrary sets (i.e., all of mathematics).

This example is a bit silly, but there is still an element of truth to it that will become apparent to anyone spending an afternoon in the library browsing through the literature of graph theory. The conceptual motivations for selecting a particular topic in graph theory are very important, especially to the beginner. The topics we shall consider in this chapter are selected because of their relationship to the study of algorithms. Even with this point of view in mind, we can only make a small selection from a very wide range of possibilities.

FIGURE 6.1 shows a labeled geometric configuration of lines (arcs, edges) and points (vertices) that represents the intuitive idea behind one standard definition of a graph.

6.1. THE INTUITIVE IDEA OF A GRAPH.

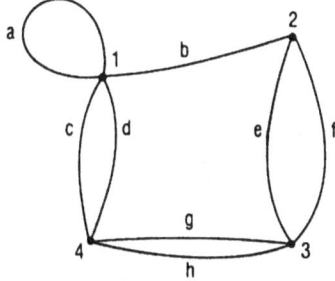

Figure 6.1

1

To describe the labeled geometric structure of FIGURE 6.1 in set theoretic terms, we can specify the *vertex* set $V = \underline{4} = \{1,2,3,4\}$, the *edge* set $\{a,b,c,d,e,f,g,h\}$ $= E$, and an *incidence* function φ from E to V. In FIGURE 6.1, the function φ is given by

$$\varphi = \begin{pmatrix} a & b & c & d & e & f & g & h \\ \{1\} & \{1,2\} & \{1,4\} & \{1,4\} & \{2,3\} & \{2,3\} & \{3,4\} & \{3,4\} \end{pmatrix}.$$

Given the triple (V,E,φ) we could draw a figure such as FIGURE 6.1. For $V = \underline{4}$ we would put four vertices on the plane (all distinct). Then, for each edge a,b,. . . of E we would draw a smooth arc joining the two vertices of $\varphi(a)$, $\varphi(b)$,. . . . The drawing need not look like FIGURE 6.1. FIGURE 6.2 shows some alternative drawings or "embeddings" of the graph (V,E,φ).

6.2 ALTERNATIVE REPRESENTATIONS OF FIGURE 6.1.

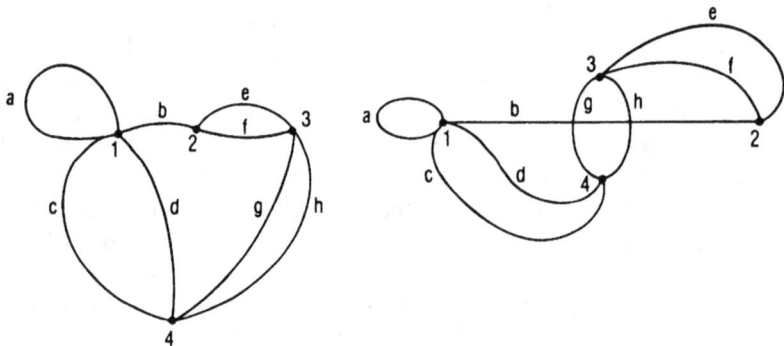

Figure 6.2

We shall take the point of view that a graph *is* a triple such as (V,E,φ). The geometrically (as point sets) different representations of FIGURES 6.1 and 6.2 all represent the same graph.

6.3 DEFINITION.

A graph $G = (V,E,\varphi)$ is a triple where V and E are arbitrary sets and φ is a function from E to subsets of V of the form $\{x,y\}$. If $x = y$ then the corresponding element of E is called a *loop* $(\{x,x\} \equiv \{x\})$. The elements of V are called the vertices of G, V(G), and the elements of E are called the edges, E(G), of G. If the function φ maps E to ordered pairs (x,y) of elements from V, then the graph is called a *directed* graph. Graphs are *equal* if their triples are equal.

6.4 PICTORIAL REPRESENTATION OF A DIRECTED GRAPH.

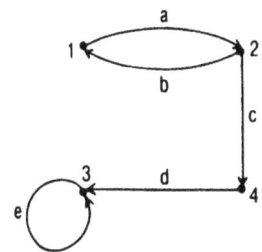

AN EDGE e IS REPRESENTED BY AN ARROW FROM x TO y IF $\varphi(e) = (x,y)$.

Figure 6.4

In many applications of graph theory a slightly simpler definition of a graph than that of DEFINITION 6.3 is used. The set E is arbitrary, so suppose E is itself a set of pairs of the form {x,y}, x and y in V. Suppose also that the function φ is the identity $\varphi(\{x,y\}) = \{x,y\}$. The resulting triple is then a graph by DEFINITION 6.3. However, the function φ is redundant, so in this case the graph may be specified by a pair, $G = (V,E)$ where E is a collection of pairs of the form {x,y}, x and y in V. FIGURE 6.5(a) shows a pictorial representation of the graph $G = (\underline{4}, \{1,2\}, \{2,3\}, \{2,4\}, \{2\})$. For graphs of the form (V,E), only one edge joins any pair of vertices in its pictorial representation. There are many other variations on the definition of a graph. One could define a *multigraph* to be a pair (V,E) where E is a *multiset* (set with repeated elements) of elements of the form {x,y}, x and y in V. FIGURE 6.5(b) represents a graph of this type with {1,2} repeated three times. One could consider instead pairs of the form (V,E) where the elements of E are arbitrary subsets of V. Such structures are called *hypergraphs*. Are either of these latter examples covered by DEFINITION 6.3?

6.5 GRAPH OF THE FORM (V,E) AND A MULTIGRAPH.

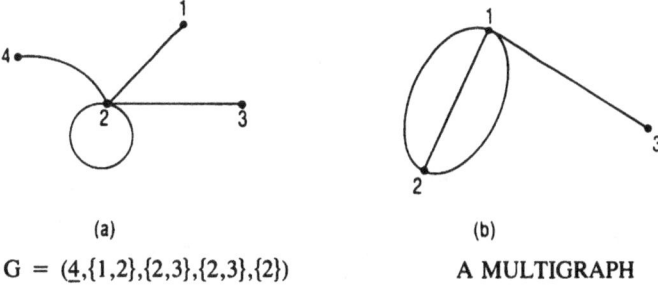

(a)

$G = (\underline{4},\{1,2\},\{2,3\},\{2,3\},\{2\})$

(b)

A MULTIGRAPH

Figure 6.5

For the most part, we shall use graphs of the form G = (V,E) and only occasionally the more general graphs specified by DEFINITION 6.3. If G = (V,E) and E is a subset of V × V, then G is *directed*.

We now introduce the idea of an ordered graph. This idea is basic to the material that follows.

6.6 DEFINITION.

Let G = (V,E,φ) be a graph and x $\in$ V. The set I_x is the set of all edges e such that x $\in$ φ(e). I_x is called the set of edges *incident on* x. If for all x $\in$ V a linear order is defined on I_x then G is called an *ordered graph*.

If the graph G is of the form (V,E), and y is such that {x,y} is an edge of G, then y is said to be *adjacent* to x. Assume G has no *loops* (pairs of the form {x,x}). Let A_x denote the set of all vertices adjacent to x. The integer $|A_x|$ is called the *degree* of x (see DEFINITION 6.14). To define a linear order on I_x it now suffices to define a linear order on A_x. Thus we have the following definition for loopless graphs of the form (V,E).

6.7 REMARK.

Let G = (V,E) be a graph without loops. If for all x $\in$ V the set A_x is linearly ordered, then G is an *ordered graph*. *Unless otherwise stated, graphs will henceforth be of the form* (V,E), *without loops*. We assume V is *finite*.

6.8 EXAMPLE.

One common method for specifying an ordered graph is with an "adjacency table." For each vertex one lists the vertex followed by an ordered list of its adjacent vertices (x:A_x).

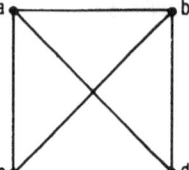

    a: c,d,b
    b: a,c,d
    c: a,b
    d: b,a

Thus, A_a consists of c,d,b in that order.

6.9 DEFINITION.

A *path of length* n in a graph is a sequence (e_1,. . .,e_n) of edges for which there exists n + 1 distinct vertices a_1,. . .,a_{n+1} such that e_i = {a_i,a_{i+1}} for i = 1,. . .,n. The sequence of vertices is called the "vertex sequence of the path." We say that the path starts at a_1 and ends at a_{n+1}.

4

6.10 DEFINITION.

A *cycle of length* n in a graph $G = (V,E)$ is a subset $\{e_1,\ldots,e_n\}$ of E such that $(e_1,\ldots,e_{n-1})$ is a path, and if the vertex sequence of this path is $a_1,\ldots,a_n$, then $e_n = \{a_1,a_n\}$. We assume $n \geq 3$.

The following diagrams represent a path and a cycle (both of which may be regarded as subgraphs of G, DEFINITION 6.43).

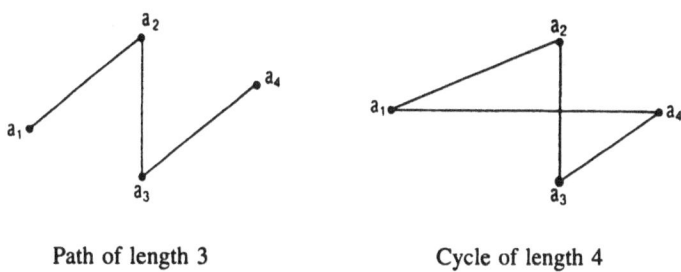

Path of length 3   Cycle of length 4

Figure 6.10

If we start with any graph and define $u \sim v$ if there is a path starting at u and ending at v, then, adopting the convention that $u \sim u$ for all u, it is easily seen that $\sim$ is an equivalence relation (reflective, symmetric, and transitive) on vertices V of G. Thus V decomposes into equivalence classes A,B, . . . under this equivalence relation. Each class A,B,. . . defines a graph by associating with that class all edges of G with both endpoints in that class (i.e., edges contained in that class as sets). The resulting graphs $G_A, G_B, \ldots$ are called the "connected components" of G. DEFINITIONS 6.9, 6.10, and 6.11 have obvious analogs for the general graph of DEFINITION 6.3.

6.11 DEFINITION.

A graph is *connected* if it has only one connected component.

6.12 DEFINITION.

A graph which is connected and acyclic is called a *tree*. Let TR(V) denote the set of all trees with vertex set V.

Here "acyclic" means "without cycles." We consider any graph with just one vertex, $|V| = 1$, a tree (the trivial tree). All graphs with two or three vertices are trees except for the cycle of length 3. In general, there are n^{n-2} trees on a vertex set V of cardinality $n \geq 2$. We give one of the many proofs of this fact below.

6.13 EXERCISE.

With V = $\underline{n}$ = {1,2,. . .,n}, list all trees with n $\leqq$ 4 vertices.

6.14 DEFINITION.

The degree of a vertex x in a general loopless graph is the number of edges incident on x (or $|A_x|$, the number of vertices adjacent to x if G = (V,E) and G is loopless).

As is obvious from the usage, an edge {x',y'} is "incident" on a vertex x if x $\in$ {x',y'} (see DEFINITION 6.6).

6.15 EXERCISE.

A vertex of degree 1 in a graph is a "*pendant*" vertex (also called an *external* or *terminal* vertex or a *leaf*). Show that every nontrivial tree has at least two pendant vertices.

In order to better understand the set of all trees with n vertices, we now consider a classical bijection between the set of all such trees and the set of functions $V^{\underline{n-2}}$ = {f: f: $\underline{n-2}$ → V}. First we give a canonical ordering of the edges of a tree. Each edge in this sequence will be a *directed edge* (i.e., an ordered pair (x,y)). Let T = (V,E), $|V| \geqslant 2$, be a tree. Let PEND(T) denote the set of pendant vertices of T. For sake of notational convenience, we assume that V is a set of nonnegative integers. Thus "x > y" means x is a larger integer than y.

6.16 *procedure* ESEQ(T).

("ESEQ" stands for *Edge Sequence*)
if $|V|$ = 2 *then* ESEQ(T): = (x,y) where V={x,y}, x > y *else*
begin
   m: = MAXPEND(T); the integer m is taken to be the maximum of the set
                      PEND(T))
   m': = the vertex adjacent to m;
   T': = (V – {m},E – {{m,m'}});
   ESEQ(T): = (m,m')ESEQ(T');
end

As an example, consider the tree T = ($\underline{4}$, {{1,3},{3,2},{3,4}}) with diagram:

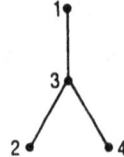

Executing the algorithm 16 for this example yields the following values:

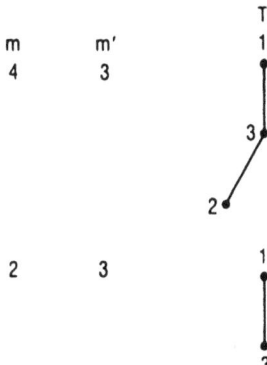

m	m'	T'
4	3	
		1
		3
		2
2	3	1
		3

Thus $ESEQ(T): = (4,3)(2,3)(3,1)$. *Note that these edges are directed.*

It is obvious that given the edge sequence $ESEQ(T)$ of a tree T, one can reconstruct the tree (simply ignore ordering and direction on the edges and one obtains immediately the pair (V,E)). It is less obvious but not difficult to see that, given only the second term of each pair in the edge sequence, one can also reconstruct the whole edge sequence if V is known.

6.17 DEFINITION.

Let T be a tree and let $ESEQ(T) = (a_1,b_1)(a_2,b_2) \ldots (a_{n-1},b_{n-1})$ be the edge sequence of T ($|V| = n$). The sequence $b_1,b_2,\ldots,b_{n-2}$ is called the *Prüffer sequence* of T.

It is useful to write a procedure for generating the Prüffer sequence directly

6.18 *procedure* PRU(T).

if $|V| = 2$ *then* PRU(T) is the empty sequence *else*
begin
   $m: = $ MAXPEND(T);
   $m': = $ vertex adjacent to m;
   $T': = (V - \{m\}, E - \{\{m,m'\}\})$;
   PRU(T): $= m'$PRU(T');
end

We now prove some basic facts about the map PRU, including the fact that PRU is a bijection between the set TR(V) of all trees with vertex set V ($|V| = $ n) and the set V^{n-2} of all functions from $\{1,\ldots,n-2\}$ to V (or, equivalently, "all sequences of length $n - 2$ with entries from V").

7

6.19 LEMMA

If $PRU(T) = b_1, \ldots, b_{n-2}$ then the set $\{b_1, \ldots, b_{n-2}\}$ is the set of all internal nodes (i.e., not pendant) of T.

Proof. The proof is by induction on $|V| = n$. If $n = 2$ then $PRU(T)$ is the empty sequence and the set $INT(T)$ of internal nodes is empty. Assume that the lemma is true for all $p < n$. If $PRU(T) = b_1, \ldots, b_{n-2}$, then it is evident from *Procedure* 6.18 that $INT(T') \cup \{b_1\} = INT(T)$. But by the induction hypothesis it follows that $INT(T') = \{b_2, \ldots, b_{n-2}\}$ since $PRU(T') = b_2, \ldots, b_{n-2}$. The lemma follows.

6.20 LEMMA.

The mapping PRU is an injection (i.e., $PRU(T) = PRU(S)$ implies $T = S$, where equality of trees means they are equal as graphs).

Proof. The proof is by induction on $|V| = n$. The result is easily verified for $n = 2,3$. Assume the lemma is true if $p < n$. Let T and S be in $TR(V)$ (DEFINITION 12) where $|V| = n$. Assume that $PRU(T) = PRU(S)$ and show that this implies that $T = S$. Following *Procedure* PRU(T), 6.18, we see immediately that $MAXPEND(T) = MAXPENDT(S) = m$ and $m' = b_1$ where $PRU(T) = PRU(S) = b_1, \ldots, b_{n-2}$. Let T' and S' represent T and S after removing the vertex m and the edge $\{m,m'\}$ from both trees. Notice that by 6.18 we have $PRU(S') = PRU(T') = b_2, \ldots, b_{n-2}$ and hence by induction $S' = T'$. Thus $S = T$.

We now give a procedure (INVPRU or "inverse Prüffer") that, given a sequence $b_1, \ldots, b_{n-2}$ in V^{n-2}, constructs a tree T such that $PRU(T) = b_1, \ldots, b_{n-2}$. The set V is given, $|V| = n \geq 2$.

6.21 *procedure* INVPRU.

$(b_1, \ldots, b_{n-2}, V)$\{Constructs $ESEQ(T)$ hence T\}
if the sequence $b_1, \ldots, b_{n-2}$ is empty $(n = |V| = 2)$
then $INVPRU(\phi, V) = (x,y)$ where $V = \{x,y\}$, $x > y$,
else
begin
   $m := MAX(V - \{b_1, \ldots, b_{n-2}\})$;
   $V' := V - \{m\}$;
   $INVPRU(b_1, \ldots, b_{n-2}, V) := (m,b_1)INVPRU(b_2, \ldots, b_{n-2}, V')$;
end

6.22 EXERCISE.

(1) Give several examples of INVPRU for $n = 8$. Give a *careful* proof that INVPRU is correct in general.

(2) Using PRU and INVPRU, give procedures for ordering trees lexicographically (CHAPTER 1, DEFINITION 1.20) in some natural fashion. (*Hint:* Consider V^{n-2}.) Describe a method for constructing a tree from TR(V) at random (such that each tree in TR(V) is equally likely to be the output of your procedure, see CHAPTER 1, EXERCISE 1.61(3) and preceding discussion).

Note that LEMMA 6.20 *and* EXERCISE 6.22(1) *prove that* PRU *is a bijection between* TR(V) *and* V^{n-2} *and hence that* $|\text{TR(V)}| = n^{n-2}$.

The notion of a tree is very useful as a descriptive tool in algorithmics. For this purpose, we shall want to add a bit more structure to the idea of a "tree."

6.23 DEFINITION.

Let $T = (V,E)$ be a tree and $v \in V$. The pair (T,v) is called a *rooted tree*. The vertex v is called the *root*. The notation RTR means rooted tree.

There are exactly $n = |V|$ possible choices for the root v and hence there are n^{n-1} possible rooted trees on a vertex set of size n. For simplicity of notation we shall often say "let T be a rooted tree" with the root v being implied or otherwise specified. The diagram of a rooted tree is usually just like that of a tree except that one vertex is circled, darkened, checked, or otherwise indicated.

6.24 DEFINITION.

Let T be an ordered tree and v a vertex of T. The pair (T,v) is called an ordered rooted tree (sometimes written ORTR).

The notion of an "ordered tree" referred to in DEFINITION 6.24 is just that of an "ordered graph" of DEFINITION 6.7 (a tree is a graph).

In any tree there is only one path joining any given pair of vertices (why?). Thus, in a rooted tree, for any vertex or edge, there is only one path starting at that vertex or edge and ending at the roots. We call the number of edges in such a path the *distance* of the given edge or vertex from the root. For the example associated with *Procedure* 6.16, the distance of vertex 4 from 1 is 2 and the distance of the edge {3,4} from 1 is also 2. If x and y are vertices of a rooted tree, then "{x,y} an edge of the tree" implies that the distance of x and the distance of y from the root are different. We use this fact in the following definition.

6.25 DEFINITION.

Let (T,v) be a rooted tree. Let {x,y} be an edge and assume without loss of generality that the distance from the root to x, D(x), is less than D(y). We call the directed edge (x,y) the *natural directed edge* associated with the undirected edge {x,y}. When each edge is given its natural direction we obtain the *natural directed tree* associated with (T,v). For each such edge (x,y), y is called a *son* of x (see also NOTATION 6.41).

In other words, the natural directed version of a rooted tree is obtained by directing each edge away from the root. See FIGURE 6.27(a).

In general, if G is a directed graph, the undirected version of G is obtained by replacing each directed edge (x,y) by the undirected edge {x,y}. We denote the resulting undirected graph by UND(G).

6.26 DEFINITION.

Let $(e_1, \ldots, e_n)$ be a sequence of edges in a directed graph G. If the same sequence of edges, without orientation, represents a path in UND(G) then we call the sequence $(e_1, \ldots, e_n)$ a chain in G. If the vertex sequence of this path in UND(G) is $a_1, \ldots, a_{n+1}$ and for each i = 1, \ldots, n, $e_i = (a_i, a_{i+1})$ then $(e_1, \ldots, e_n)$ is a *directed path* or simply *path* in G. In FIGURE 6.27 ((a,b),(a,c)) is a chain, ((a,b)(b,e)) a path.

In other words, *in the diagram of a directed path, all arrows point in the direction of the terminal vertex of the path.*

6.27 AN ORDERED ROOTED TREE ORTR AND DEPTH FIRST SEQUENCES.

(The order on sons, DEFINITION 6.25, is left to right.)

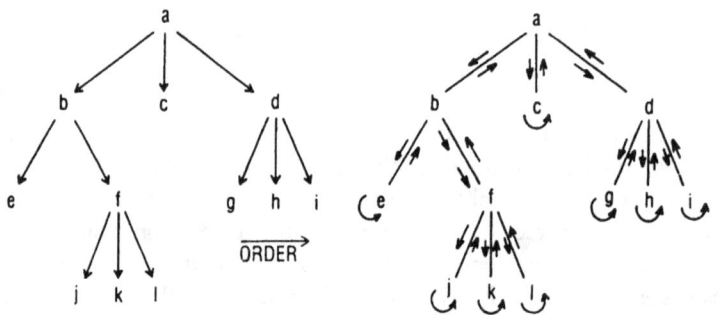

(a) An ORTR with root a (natural directed version shown for each edge).

(b) Diagram of depth first sequences. DFV(T) is abebfjfkflfbacadgdhdida.

Figure 6.27

Let (T,v) be a rooted tree where T = (V,E). Let A_v be the set of vertices adjacent to the root v. Direct T according to DEFINITION 6.25.

10

6.28 DEFINITION.

For x in A_v, define $T_x = (V_x, E_x)$ where V_x is the set of all vertices y for which there is a directed path from x to y (we include x in V_x) and E_x is the set of all edges e such that e = {a,b} is contained in V_x. The graphs of the form T_x for x in A_v are called the *principal subtrees* of the root v. The root of T_x is x.

The reader should verify that for x and y in A_v with x $\neq$ y the sets E_x and E_y are disjoint, V_x and V_y are disjoint, and T_x is in fact a tree. If T is an ordered tree, and hence A_v is ordered, then the set of principal subtrees is also ordered by the natural order induced from A_v. In diagrams, we draw the principal subtrees below the root in order left to right.

We now consider some additional ideas involving ordered rooted trees that provide very useful descriptive tools in combinatorics and algorithmics. Consider the diagram (b) of FIGURE 6.27. By following the arrows as indicated and traversing the edges of the tree, we encounter the vertices in the following order: abebfjfkflfbacadgdhdida. We call this sequence the *depth first sequence* of vertices of T, DFV(T). Each vertex of T appears in DFV(T) as many times as its degree in T except for the root which appears once more than its degree.

6.29 EXERCISE.

For any graph, let DEG(v) denote the degree of the vertex v. Show that the sum over all v of DEG(v) is twice the number of edges of G. Thus the length of DFV(T) is twice the number of edges of T plus one (2|E| + 1). Prove that in any tree T = (V,E), |E| = |V| − 1 and thus the length of DFV(T) is also 2|V| − 1.

If we again follow the arrows of the second diagram of FIGURE 6.27(b) and record an edge each time it is traversed, we obtain DFE(T), the depth first sequence of edges: {a,b} {b,e} {b,e} {b,f} {f,j} {f,j} {f,k} {f,k} {f,l} {f,l} {b,f} {a,b} {a,c} {a,c} {a,d} {d,g} {d,g} {d,h} {d,h} {d,i} {d,i} {a,d}. The sequence consists of edges of the undirected ordered rooted tree. Clearly, the length of the depth first sequence of edges, DFE(T), is 2 |E|.

Both DFV(T) and DFE(T) have natural recursive descriptions. Let T be an ordered rooted tree with principal subtrees $T_1, T_2, . . ., T_p$ in order. Let v be the root of T and v_i the root of T_i for each i.

6.30 DEFINITION.

Let T = (V,E) be an ordered rooted tree with root v. If |V| = 1 then DFV(T) = v and DFE(T) is empty. Otherwise, let $T_1, . . ., T_p$ be the principal subtrees of v in order with roots $v_1, . . ., v_p$. Then we define

11

$$DFV(T) = vDFV(T_1)vDFV(T_2)v. \ . \ .vDFV(T_p)v$$

and

$$DFE(T) = \{v,v_1\}DFE(T_1)\{v,v_1\}. \ . \ .\{v,v_p\}DGE(T_p)\{v,v_p\}.$$

Using the sequences DFV and DFE we can define two important linear orders on the vertices and edges of an ordered rooted tree.

6.31 DEFINITION.

The sequence of first occurrences of elements of V or E in the sequences DFV(T) or DFE(T), respectively, are all called the *preorder sequences* of vertices or edges. We write PREV(T) or PREE(T) for these sequences. Likewise, we define the *postorder sequences* of vertices or edges to be the corresponding last occurrences. We write POSV(T) or POSE(T) for these sequences.

For example, consider the tree T of FIGURE 6.27. The sequence PREV(T) is abefjklcdghi. The sequence PREE(T) is $\{a,b\}$ $\{b,e\}$ $\{b,f\}$ $\{f,j\}$ $\{f,k\}$ $\{f,l\}$ $\{a,c\}$ $\{a,d\}$ $\{d,g\}$ $\{d,h\}$ $\{d,i\}$. The sequence POSV(T) is ejklfbcghida. Notice that these sequences actually define linear orders on the vertices or edges of T since each vertex or edge occurs only once in the sequence. If T′ is T with the reverse order, then PREV(T′) is POSV(T) read backwards. To practice writing down the vertices of a tree in preorder and postorder, the reader should follow the arrows in the second diagram of FIGURE 6.27. Writing down a vertex the last time it is encountered (i.e., when arriving at that vertex from its rightmost son) produces postorder on vertices.

Recall that we have used the notation A_x to indicate the vertices adjacent to a vertex x in a graph. Similarly, we have used I_x to indicate the set of edges incident on x (DEFINITION 6.6).

We have already observed that between any two vertices in a tree there is one and only one path. We use that idea in the following.

6.32 DEFINITION.

Let T be a rooted tree with root v. For each edge e of T we define the stack of e, STACK(e), to be the path starting at v and ending with e. Similarly, for each vertex x ≠ v, define STACK(x). Define STACKS(T) to be the set {STACK(e): e an edge of T}.

We now define a natural version of lexicographic order ("length first lex order" CHAPTER 1, EXERCISE 1.29) on the set STACKS(T). Let T be ordered, and for each vertex x let I_x be the incident edges as above.

6.33 DEFINITION.

(BREADTH FIRST ORDER) Let $e = (e_1,. \ . \ .,e_p)$ and $e' = (e_1',. \ . \ .,e_q')$ be two paths in STACKS(T) where T is an ordered rooted tree. Define $e < e'$ if either

$p < q$ or $p = q$ and the smallest t such that $e_t \neq e_t'$ satisfies $e_t < e_t'$. *Note:* e_t and e_t' are, because of the definition of t, both incident on a common vertex x. Thus they are ordered by the ordering on I_x (or equivalently A_x) given by the ordering on the tree T. This is the meaning of "$e_t < e_t'$."

The STACK of DEFINITION 6.32 corresponds to the stack of local data maintained by any implementation of a recursive program. The mapping STACK from E to STACKS(T) is clearly a bijection. Thus the breadth first order on STACKS(T) defines a linear order on E called breadth first order on edges, BRE(T). Similarly, we define breadth first order on vertices, BRV(T). In the latter case we add the root v to the order and adopt the convention that it is the first element. If the tree T is drawn as in FIGURE 6.27, then breadth-first order on vertices is obtained by listing the root, then all vertices of distance one from the root left to right, then all of distance two left to right, etc. In this example breadth first order is just alphabetic order: abcdefghijkl.

6.34 EXERCISE.

Given just the preorder and postorder sequences of vertices of a tree, can one reconstruct the tree? Given breadth first order and preorder or breadth first order and postorder can one reconstruct the tree?

We now discuss informally the relationship between ordered rooted trees and recursively described algorithms. As an example, consider the classical Towers of Hanoi puzzle. As in FIGURE 6.35, we start with three positions A, B, and C. The stack of n discs of different size is placed initially at position A. The discs range in size from smallest to largest going from top to bottom. The problem is to transfer the discs from position A to position C according to the following rules:

(1) The discs are to be transferred one at a time and only the top disc of a stack can be moved.
(2) The disc that moved can be placed at a position with no discs or at the top of a stack at another position provided that the top disc of the stack is larger than the disc that was moved.

The sequence of moves is completed when all of the discs are transferred to position C.

The basic recursive description of the solution of the Towers of Hanoi puzzle is indicated in FIGURE 6.36.

6.35 TOWERS OF HANOI.

Let H(A,B,C,n) be the sequence of moves needed to transfer n discs from position A to position C using B for temporary placement.

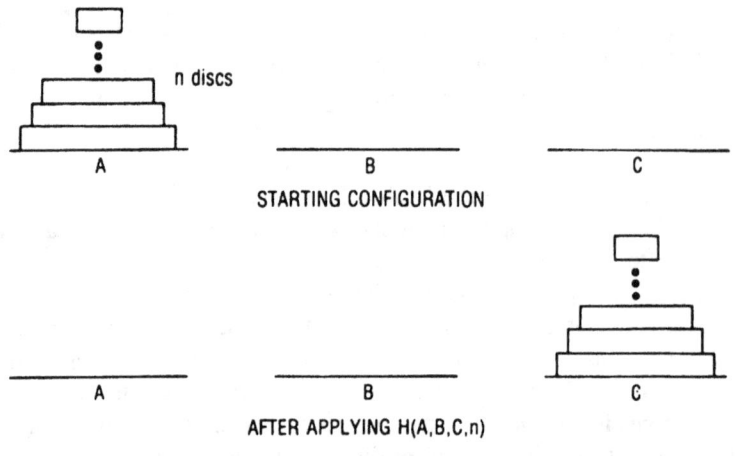

STARTING CONFIGURATION

AFTER APPLYING H(A,B,C,n)

Figure 6.35

6.36 THE BASIC RECURSION FOR THE TOWERS OF HANOI PUZZLE.

H(X,Y,Z,k) With starting configuration as in FIGURE 35 (A: = X, B: = Y, C: = Z, n: = k). *Apply* H(X,Z,Y,k − 1):

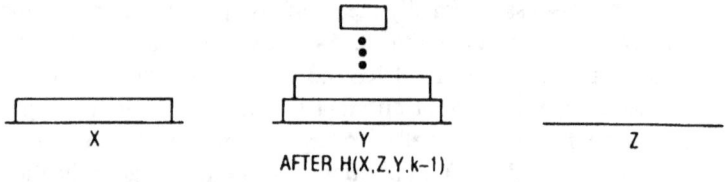

AFTER H(X,Z,Y,k−1)

Figure 6.36a

Now transfer the disc on X to Z (write "X to Z")

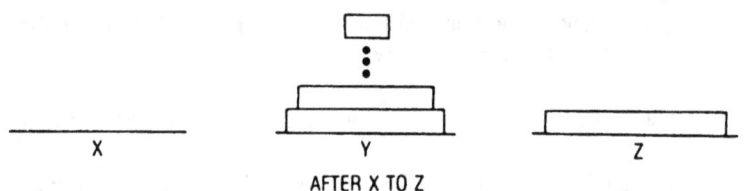

AFTER X TO Z

Apply H(Y,X,Z,k − 1):

Figure 6.36b

14

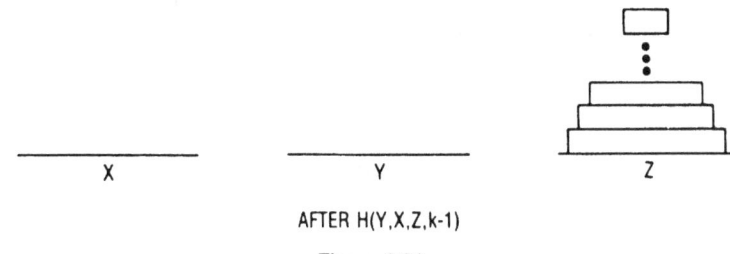

AFTER H(Y,X,Z,k-1)

Figure 6.36c

The steps of FIGURE 6.36 may be described in pidgin Algol as follows.

6.37 *procedure* **H(X,Y,Z,k).**

if k = 1 *then* write ''X to Z'' *else*
begin
   H(X,Z,Y,k − 1);
   write ''X to Z'';
   H(Y,X,Z,k − 1);
end

A program written for a computer that compiled algol and that is basically like *procedure* 6.37 would actually list the moves to solve the puzzle for various assignments of X,Y,Z, and k. We wish to have some working intuitive conventions for doing the same thing as the algol compiler.

We shall regard recursions such as described in 6.36 or 6.37 as essentially rules for constructing ordered rooted trees. In this case the rule tells us what the tree looks like ''locally.'' The diagram of this ''local'' description is given by 6.38.

6.38 LOCAL DESCRIPTION OF THE TOWERS OF HANOI TREE.

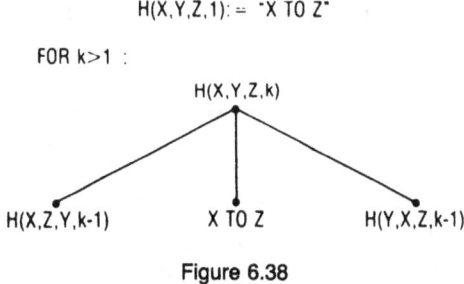

Figure 6.38

To generate a particular tree using the rules of FIGURE 6.38, one starts with the root, say H(A,B,C,n), and constructs the tree using the rule of FIGURE 6.38

15

repeatedly. Usually the tree's vertices are constructed in preorder. H(A,B,C,3) is shown in FIGURE 6.39.

6.39   H(A,B,C,3).

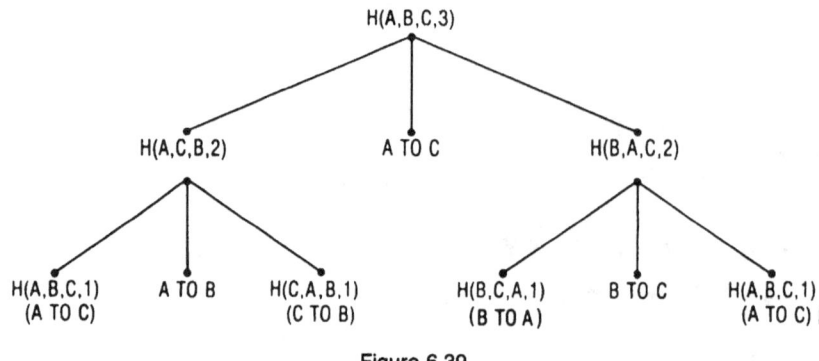

Figure 6.39

Note that we are using the symbols of the form H(X,Y,Z,k) in two ways. They are being used to stand for sequences of moves in the puzzle and as vertices for the tree of FIGURE 6.38 and FIGURE 6.39. The sequence of moves that solves the Towers of Hanoi problem H(A,B,C,3) is obtained by listing the "x to y" type statements in the tree of FIGURE 6.39 as they are encountered in a depth first listing of the vertices. (This is the order obtained by projecting the pendant vertices, or "leaves," of the tree of FIGURE 6.39 onto a horizontal line and reading from left to right. These statements of the form "x to y" form a subsequence of the preorder sequence of the tree. The reader should practice generating the tree of FIGURE 6.39 by pencil and paper using only the rules of FIGURE 6.38. Observe that if the tree is generated in depth first order (the order in which the vertices are written down is preorder in vertices) and each statement of the form "x to y" is written down in a separate list as it is generated, then one never needs to keep on paper any more of the tree than the stack of the last vertex generated (as in DEFINITION 6.32, the stack is the path back to the root).

This observation is important for complex computations when one does not wish to store the whole tree of the recursion, such as was done in FIGURE 6.39. In the case of FIGURE 6.39, the list of executable statements in preorder is A to C, A to B, C to B, A to C, B to A, B to C, A to C. The reader should check that these moves do solve the puzzle. The recursive method that we just discussed is often very useful for describing algorithms and proving theorems about combinatorics and algorithms. Often, however, the generation of the tree such as FIGURE 6.39 from the recursive description is not the most "efficient" method for generating the required output. This is certainly the case with the Towers of Hanoi problem. The geometric insight gained by thinking about the structure of

16

the "recursion tree" of various recursions can often provide valuable clues for designing more efficient "nonrecursive" methods (see EXERCISE 6.40(3)).

The following exercises contain many beautiful and entertaining surprises!

6.40 EXERCISE.

(1) Using FIGURE 6.38, generate the moves of H(A,B,C,4) by constructing the vertices of the tree like the one shown in FIGURE 6.39. For each vertex, keep only its stack. The vertices should be generated in depth first order (i.e., in preorder as if making a list of the depth first sequence of vertices). How many moves are there in H(A,B,C,n)?

(2) For n = 20, what would be the 500,000th move? If someone abandons a Towers of Hanoi problem somewhere before completion and you find only the configuration of discs left behind, how do you find the next move?

(3) Notice that in H(A,B,C,n) the smallest disc is moved every other move. The alternate moves are forced by the rule that a disc cannot be placed on top of a smaller one. Determine the pattern of moves of the smallest disc (it depends on whether n is odd or even) and use this to give a nonrecursive rule for constructing the sequence of moves H(A,B,C,n). Give a rigorous proof that your rule works (for the proof you will probably want to go back to the recursion!).

(4) Suppose that instead of having only three places A, B, and C to stack discs, there are four places A, B, C, and D. Study H(A,B,C,D,n).

(5) Let $\Pi(n,k)$ denote the set of all partitions of the set $\underline{n}$ into k blocks (see CHAPTER I, DEFINITION 1.3). If $\alpha \in \Pi(n,k)$ is a partition, then α is, by definition, an unordered collection of subsets of $\underline{n}$. For the sake of discussion, order the blocks of any partition such as α by the smallest element in each block. Thus, $\alpha = (\{1,7,8\}, \{2,3,6\}, \{4,5\})$ is an element of $\Pi(8,3)$ ordered according to the above convention. Note that in this example, α may be thought of as having been obtained from $\beta \in \Pi(7,3)$, $\beta = (\{1,7\}, \{2,3,6\}, \{4,5\})$, by adding 8 to the first block of β. We write $\alpha = (8 \rightarrow 1)\beta$. For $\gamma = (\{1,7\}, \{2,3,6\}, \{4,5\}, \{8\})$ we write $\gamma = (8 \rightarrow \infty)\beta$. The following diagram summarizes this situation:

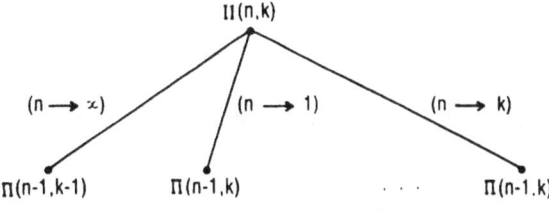

Figure 6.40

The diagram states that $\Pi(n,k)$ can be constructed by first applying the operation $(n \rightarrow \infty)$ to all partitions in $\Pi(n-1,k-1)$, then applying $(n \rightarrow 1)$

17

to all partitions in $\Pi(n-1,k)$, then $(n \rightarrow 2)$ to all partitions in $\Pi(n-1,k)$, etc. Note that this diagram is analogous to FIGURE 6.38 for $H(A,B,C,n)$. Thus this diagram is the local description of a tree. For particular n and k, study this tree. Use these ideas to develop a method for listing the partitions $\Pi(n,k)$. The numbers $S(n,k) = |\Pi(n,k)|$ are called the *Stirling numbers of the second kind*. Note that the above diagram implies that $S(n,k) = S(n-1,k-1) + kS(n-1,k)$. Use this recursion to make a table of the $S(n,k)$. (The reader may wish to review FIGURES 1.42, 1.48, and 1.52 of CHAPTER 1 for related ideas.)

(6) The Towers of Hanoi problem involved two recursive calls. Here is an interesting example that involves three recursive calls. This example and Problem (7) are a bit more complex than those we have been considering. Problem (7) involves a little group theory. Consider k 1's and $n - k$ 0's in an array F with $F(1) = \ldots = F(k) = 1$. We wish to make a series of single interchanges of 0's and 1's in the array F such that all k subsets of $\underline{n} = 1,2,\ldots,n$ are at some point represented in F and such that the final configuration is $F(n - k + 1) = \ldots = F(n) = 1$. If n = k we do nothing. For k = 1 the sequence is $1000\ldots00, 0100\ldots00, 0010\ldots00, \ldots,$ $0000\ldots10,0000\ldots01$. Recursively we may specify the procedure by first moving, using the procedure, the k ones from their initial position to $F(n - k) = \ldots = F(n - 1) = 1$ thus generating all subsets of size k with $F(n) = 0$. We then set $F(n - 1) = 0$ and $F(n) = 1$. Now run the procedure in reverse moving the $k - 1$ 1's to $F(1) = \ldots = F(k - 1) = 1$ generating all k subsets of $\underline{n}$ with $F(n - 1) = 0$, $F(n) = 1$. Next set $F(k - 1) = 0$ and $F(n - 1) = 1$. Finally apply the procedure to the $k - 2$ 1's specified by $F(1) = \ldots = F(k - 2) = 1$ ending with $F(n - k + 1) = \ldots = F(n - 2) = 1$ and generating all k-subsets of $\underline{n}$ with $F(n - 1) = F(n) = 1$. The final configuration is the desired one and all subsets of size k have been generated.

We now give a formal definition of the procedure. Let $F(1),\ldots,F(n)$ be an array with $F(i) = 0$ or 1. Let $\alpha,\beta \in \underline{n}$. Suppose t (representing the number of 1's) is an integer, $0 \leq t \leq |\alpha - \beta| + 1$ and let $\delta = +1$ or -1. The procedure MOVE(t,δ,α,β) is defined if $\delta = +1$ and $\alpha \leq \beta$ or $\delta = -1$ and $\alpha \geq \beta$. MOVE(t,δ,α,β) generates sequentially in locations $F(\alpha)$, $F(\alpha+\delta),\ldots,F(\beta)$ all 0,1 sequences with t ones. Initially, if $t \geq 1$, $F(\alpha) = F(\alpha+\delta) = \ldots = F(\alpha+(t-1)\delta) = 1$ and finally $F(\beta) = F(\beta-\delta) = \ldots = F(\beta-(t-1)\delta) = 1$.

procedure MOVE(t,δ,α,β) {will generate sequentially in locations $F(\alpha)$,
$F(\alpha+\delta),\ldots,F(\beta)$ all 0,1 sequences with t 1's}
begin
    if $t = 0$ or $t-1 = |\alpha - \beta|$ *do* return;
    if $t = 1$ shift the 1 one step at a time from $F(\alpha)$ to $F(\beta)$;
    else {at this point $2 \leq t \leq |\alpha - \beta|$

begin

MOVE $(t,\delta,\alpha,\beta-\delta)$; {ends with $F(\beta-t\delta) = \ldots = F(\beta-\delta)$
$= 1$}

$F(\beta-\delta):=0$ and $F(\beta):=1$; {switch a 0,1 pair}

MOVE $(t-1,-\delta,\beta-2\delta,\alpha)$; {ends with $F(\alpha) = \ldots$
$= F(\alpha+(t-2)\delta) = 1.$
$F(\beta-\delta) = 0$ and $F(\beta) =$
1 throughout}

$F(\alpha+(t-2)\delta):=0$ and $F(\beta-\delta):=1$; {switch a 0,1 pair}

MOVE $(t-2,\delta,\alpha,\beta-2\delta)$; {ends with
$F(\beta)=\ldots=F(\beta-(t-1)\delta) = 1$}.

end

Execute MOVE $(3,1,1,6)$ and some other examples to get a feeling for the procedure. Given a 0,1 sequence, be able to find the next sequence in MOVE $(t,1,1,n)$. If you have studied PART I, try and find good algorithms for computing RANK and UNRANK for the list of sequences in MOVE $(t,1,1,n)$.

(7) As usual, let S_n denote all permutations of $\underline{n}$. Let C_n denote the cyclic group generated by the permutation $\tau = (12\ldots n)$. C_n acts on S_n by the rule $\tau*\sigma(i) = \sigma(\tau^{-1}(i))$. The standard "direct insertion" method (see CHAPTER 1, FIGURE 1.47) for recursively generating S_n proceeds by selecting in order (recursively) an element from S_{n-1}, $\sigma = \sigma_1,\sigma_2,\ldots,\alpha_{n-1}$, and inserting n in the various possible positions:

$$\sigma_1\sigma_2\ldots\sigma_{n-1}n,\ \sigma_1\sigma_2\ldots n\sigma_{n-1},\ldots,\sigma_1 n\sigma_2\ldots\sigma_{n-1},\ n\sigma_1\sigma_2\ldots\sigma_{n-1}.$$

Given any order on S_{n-1} we could use this procedure, denoted by $S_{n-1}(n)$ say, to form S_n. One possibility is to decompose S_{n-1} into orbits under the action of C_{n-1}: $S_{n-1} = O_1 \cup O_2 \cup \ldots \cup O_p$, $p = (n-2)!$ Order the orbits in some manner and the elements in the orbits in some manner and form $S_{n-1}(n) = O_1(n) \cup \ldots \cup O_p(n)$. It is well known that a system of orbit representatives for C_{n-1} acting on S_{n-1} is simply $\{\varphi: \varphi \in S_{n-1}, \varphi(n-1) = n-1\}$. Thus the orbits are identified with S_{n-2}. With a little care one can then proceed recursively. Consider $\sigma_1\sigma_2\ldots\sigma_{n-1}n$. After the series of insertions of n we obtain $n\sigma_1\sigma_2\ldots\sigma_{n-1}$. Switching the first and last symbols gives $\sigma_{n-1}\sigma_1\ldots\sigma_{n-2}n$ which is just $\tau*(\sigma_1\ldots\sigma_{n-1})n$ where $\tau = (12\ldots(n-1))$ is the generator of C_{n-1}. Thus $n-1$ series of n insertions followed by a switch of the first and last elements produces $O_t(n)$ where O_t is the orbit of C_{n-1} containing $\sigma_1\ldots\sigma_{n-1}$ (and returns us to the starting permutation $\sigma_1\sigma_2\ldots\sigma_{n-1}n$). Thus starting with 12345 say, we would go 12345, $\ldots$, 51234, 41235, $\ldots$ 54123, 34125, $\ldots$, 53412, 23415, $\ldots$, 52341. Thus far we have formed $O_1(5)$ where O_1 is the orbit of C_4 acting on S_4 which contains 1234. The next switch brings us back to 12345 at which point we apply the procedure recursively to 123 to get the representative 1324 of the next orbit of C_4 acting on S_4. It should be obvious from the above remarks about the action of C_n on S_n that the algorithm works in general.

We give a rough pidgin algol description of the above procedure. Suppose an array A(1). . .A(n) contains a permutation of $\underline{m}$ in positions A(1). . .A(m). We define a procedure NEXTPERM(A,m) which updates the array A(1),. . .,A(n). We suppose that a procedure I(A,m) returns the location of m (if A ← 12534 then I(A,5) = 3 for example) and a procedure T(A,m − 1) deletes m and returns the relative position of m − 1 mod (m − 1) (if A← 12534 then T(A,4) = 0, if A ← 54123 then T(A,4) = 1, if A ← 34125 then T(A,4) = 2, if A ← 53142 then T(A,4) = 3). Show that the following procedure may be used to generate S_n:

Procedure NEXTPERM(A,m) {To generate S_n start with A(i) = i, i = 1,. . .,n, and set m: = n. Apply NEXTPERM(A,n) over and over again until LAST occurs}

```
begin
  if m = 1 write LAST;
  if m = 2 then if A(1) = 1 then
  begin
    SWITCH(A(1), A(2));
    write A(1),. . .,A(n)
  end
  else write LAST;
  if I(A,m) > 1 then
    begin
      SWITCH (A(I(A,m)),A(I(A,m) − 1))
      write A(1),. . .,A(n)
    end
  else
  if T(A,m − 1) ≠ m − 2 then { in A = (5,2,3,4,1,. . .) with m = 5
                                T(A,4) = m − 2 = 3}
    begin
      SWITCH (A(1),A(m))
      write A(1),. . .,A(n)
    end
  else
    begin
      SWITCH (A(1), A(m));
      NEXTPERM (A,m − 2)
    end
end
```

Generate the first 40 elements of NEXTPERM (A,5). By examples, show how to generate the predecessor and successor of selected permutations for n = 6,8. Show how to compute RANK and UNRANK for NEXTPERM.

Note. 1234, 1243, 1423, 4123, 3124, 3142, 3412, 4312, 2314, 2341, 2431, 4231, 2134, 2143, 2413, 4213, 3214, 3241, 3421, 4321, 1324, 1342, 1432, 4132, LAST is the result of applying NEXTPERM(A,4) 24 times starting with A = (1,2,3,4).

(8) Sometimes two or more recursive procedures are defined together as the following basic example illustrates. As usual, let $\{0,1\}^{\underline{n}}$ denote all functions from $\underline{n}$ to $\{0,1\}$ or, equivalently, all sequences of length n formed from the symbols 0 and 1 (i.e., using one-line notation for functions). We use $\overrightarrow{\text{GRAY}}(n)$ to denote a certain linear order on these sequences and $\overleftarrow{\text{GRAY}}(n)$ to denote this same linear order written backwards. To start with, $\overrightarrow{\text{GRAY}}(1)$ = 0,1 so $\overleftarrow{\text{GRAY}}(1)$ = 1,0. The notation $1\overrightarrow{\text{GRAY}}(n-1)$ means "adjoin 1 to the front of each string in $\overrightarrow{\text{GRAY}}(n-1)$." Thus, $1\overrightarrow{\text{GRAY}}(1)$ = 10,11. Similarly, define $0\overrightarrow{\text{GRAY}}(n-1)$, $1\overleftarrow{\text{GRAY}}(n-1)$, and $0\overleftarrow{\text{GRAY}}(n-1)$. We define $\overrightarrow{\text{GRAY}}(n)$ = $0\overrightarrow{\text{GRAY}}(n-1)$, $1\overleftarrow{\text{GRAY}}(n-1)$ and $\overleftarrow{\text{GRAY}}(n)$ = $1\overrightarrow{\text{GRAY}}(n-1)$, $0\overleftarrow{\text{GRAY}}(n-1)$. (By introducing a parameter indicating direction, these two procedures can obviously be written as one; this can be done in general for this type of recursive description.) Some examples: $\overrightarrow{\text{GRAY}}(2)$ = $0\overrightarrow{\text{GRAY}}(1)$, $1\overleftarrow{\text{GRAY}}(1)$ = 00,01,11,10. $\overrightarrow{\text{GRAY}}(3)$ = $0\overrightarrow{\text{GRAY}}(2)$, $1\overleftarrow{\text{GRAY}}(2)$ = 000,001,011,010,110,111,101,100. Prove that the sequences in the list GRAY(n) differ from their successor in exactly one coordinate. Analyze the list $\overrightarrow{\text{GRAY}}(n)$ in the same manner that the above recursions have been analyzed. (Given m, what is the mth entry in the list? Given a sequence, what is its predecessor and successor? etc.) The sequence $\overrightarrow{\text{GRAY}}(n)$ is called the "binary Gray code."

end EXERCISE

Before continuing our discussion of trees as they relate to the study of more general graphs (as spanning trees), it will be worthwhile to introduce some additional notation.

6.41 NOTATION FOR TREES.

Let T be a rooted tree with root v. Regard T as a directed tree in the natural manner (each edge directed away from the root as in DEFINITION 6.25). If (x,y) is a directed edge, we say that x is the *father* of y and y is the *son* of x. If x and y are vertices connected by a directed path from x to y, then y *is a descendant of* x and x *is an ancestor of* y. In this case we say that x and y are *lineal descendants* or a *lineal descendant* pair. If e = {x,y} is an edge with corresponding directed edge (x,y) then we call x the initial vertex of e (we write x = IN(e)) and y the terminal vertex (TM(e)) of e.

6.42 NOTATION FOR VECTORS.

Let L = (a,b,. . .,e,f) be a vector or equivalently, a list, sequence, or linearly ordered set ab. . .ef. We use the following notations: $\leftarrow$ L stands for the statement L: = (b,. . .,e,f); x $\leftarrow$ L stands for the two statements x: = a and $\leftarrow$ L; x $\rightarrow$ L stands for L: = (x,a,b,. . .,e,f). Similarly, we define L $\rightarrow$, L $\rightarrow$ x, and L $\leftarrow$ x.

6.43 DEFINITION.

Let G = (V,E) be a graph. A graph G' = (V',E') satisfying the conditions V' $\subseteq$ V and E' $\subseteq$ E will be called a *subgraph* of G (written G' $\subseteq$ G).

We now give another description of how a recursive algorithm can be described using ordered rooted trees. If the Towers of Hanoi problem is solved by constructing the tree specified by the local description FIGURE 6.38 and if the tree is constructed in depth first order (i.e., the vertices or edges are constructed as they would occur in the depth first sequence of edges or vertices of the tree) then a sequence of trees $T_1, T_2, . . ., T_p$ is constructed. Each $T_i \subset T_{i+1}$ and T_p is the final tree (as shown in FIGURE 6.39 for n = 3, for example). We have noticed in EXERCISE 6.40(1) that in order to generate T_{i+1} from T_i we need only know the stack of the preorder last edge of T_i. To obtain T_{i+1} (in Towers of Hanoi) we start with the last edge in the stack (in this case, the preorder last edge of T_i) and go backwards through the stack to find the first edge (x,y) with x of degree 1 or 2 in T_i. The edge (x,y'), y' the next son of x after y, is then added to T_i.

This idea of constructing a sequence of trees gives a useful geometric method for describing a wide class of algorithms. We give a general description of this process. Let T_i be a tree and let S be the stack of the preorder last element of T_i (T_i is ordered and rooted). We assume that S is the vertex sequence of the stack. Thus, the sequence S = (1,3,5,4) would represent the sequence of (directed) edges ((1,3),(3,5),(5,4)) in the tree of FIGURE 6.44 where 4 is the preorder last element.

22

6.44 STACK OF PREORDER LAST ELEMENT.

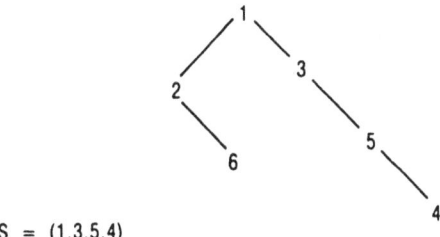

S = (1,3,5,4)

Figure 6.44

We use the notation $x < y$ to indicate that x is closer to the root than y in S. We shall construct a sequence of trees $T_1, . . ., T_p$ inductively by adding one edge at a time. Consider a tree T_i (all are ordered rooted trees with the same root v). Let S be the stack of the preorder last element w. Suppose we have a function or "rule" for deciding whether or not any $x \in S$ has a son in T_p that is not in the tree T_i. We call this function SON and write $SON(x,T_i) = \phi$ if x does not have a son in $T_p - T_i$. If x does have a son in $T_p - T_i$ we assume that $SON(x,T_i)$ is the first such son. Repeated applications of SON to the elements of S, starting with the last element of S and going in reverse order will either produce this latter case or find that $SON(x,T_i) = \phi$ for all x in S. This is the idea behind *procedure* 6.45.

In the Towers of Hanoi problem, the rule SON is specified by the local diagram of the tree given by FIGURE 6.38. The *procedure* 6.45 gives a straightforward description of how the trees of the Towers of Hanoi problem were constructed but in a slightly more general setting. Initially, we take $T_1 = (\{v\},\phi)$ and S = (v). By repeated applications of the following procedure, we construct a sequence of trees $T_1, . . ., T_p$. In each case, S is initially the stack of the preorder last element of T_i. Recall NOTATION 6.42.

6.45 *procedure* NEXT(T_i).

while $SON(LAST(S),T_i) = \phi$ *do* S→; {find vertex of S that has a son}
if S = ϕ exit *else*
begin
   (x,y): = (LAST(S),SON(LAST(S),T_i));
   NEXT(T_i): = T_i with {x,y} added; {NEXT(T_i) is the tree T_{i+1}}
   S ← y;
end

In each tree T_i, the root is v and the order is determined by the order in which the edges incident to a given vertex are added. The procedure NEXT represents the basic idea involved in converting a recursive implementation of an algorithm into a nonrecursive implementation by use of a stack.

6.46   EXAMPLE.

We consider an example of *procedure* 6.45 applied to the Towers of Hanoi problem. Consider the case n = 6 and the stack H(A,B,C,6), H(B,A,C,5), H(B,C,A,4), H(C,B,A,3), H(B,C,A,2), H(C,B,A,1). Reading this stack from right to left and referring to FIGURE 6.38, we see that the largest element in the stack to have another son in the full tree is H(B,A,C,5) and the first such son is "B to C." Thus we add the edge (H,(B,A,C,5), "B to C") to the tree T_i (i.e., the existing tree, of which we actually know only the stack at the moment) and obtain the new stack: H(A,B,C,6), H(B,A,C,5), "B to C." Notice that in this example, the function SON depends on the stack of T_i and the local description of the tree T_p (the final tree). This is often the case. We need only the recursive description and the stack.

We now consider another example of an algorithm that involves the generation of a sequence of trees by successive augmentations of the stacks of the trees. It is necessary to first introduce the important notion of a spanning tree of a graph.

6.47   DEFINITION.

If G = (V,E) is a connected graph and G' = (V',E') is a subgraph with V' = V, then G' is called a *spanning subgraph*. If, in addition, G' is a tree, then G' is called a *spanning tree for* G. A graph whose components are trees is called a *forest*. For any graph G, a subgraph G' whose **components** are spanning trees for the **components** of G is called a *spanning forest* for G. If G' is a spanning forest of G, then the edges of E − E' are called the *chords* of G' in G. A pair (G', v), v ∈ V and G' a spanning tree, is called a *rooted* spanning tree. The vertex v is called *the root* of (G', v).

6.48   EXERCISE.

Give a careful proof that any connected graph G must have at least one spanning tree. *Hint*: Show that if G' = (V',E') is a subgraph of G, and G' is a tree (we call G' a *subtree* in this case), then V' not equal to V implies that a larger subtree of G can be constructed from G' by adding one more edge.

6.49   DEFINITION.

Let G = (V,E) be a graph and let (G', v) be a rooted spanning tree for G. If for every edge {x,y} of G, the vertices x and y are a lineal descendant pair with respect to the rooted tree G' then we say that G' is a *lineal spanning tree* of G (recall Notation 6.41).

If the edge {x,y} of DEFINITION 6.49 is an edge of G', then it is obvious that x and y are a lineal descendant pair. Thus the condition must be checked only on the edges of G that are not in G'. Consider EXAMPLE 6.50.

6.50 EXAMPLE.

Two rooted (at *a*) spanning trees of G, natural directions shown.

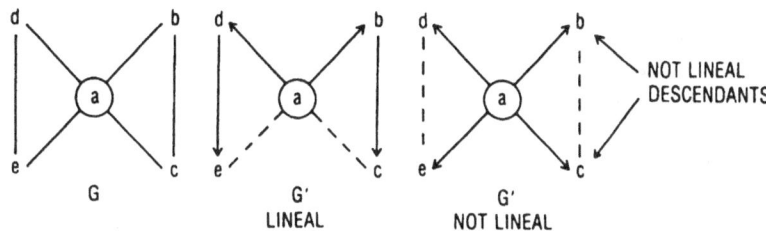

Figure 6.50

In EXAMPLE 50, the edges {b,c} and {d,e} both have vertices that do not form a lineal descendant pair with respect to the second spanning tree.

We shall now show that given any connected graph G and any vertex v of G, there is a lineal spanning tree of G rooted at v. To do so, it is convenient to extend the notion of "lineal" to subtrees of G that are not necessarily spanning. Let V(G) and E(G) denote the vertices and edges of G.

6.51 DEFINITION.

Let G be a graph and T a subtree of G rooted at v. We say that T is a lineal subtree of G if T is *ordered* and satisfies the following:

(1) If e = {x,y} is an edge of G (e ∈ E(G)) with both endpoints vertices of T (e ⊆ V(T)) then these endpoints x and y are a lineal descendant pair with respect to T.
(2) If e ∈ E(G) has exactly one vertex in common with T, then that vertex is in STACK(e') where e' is the preorder last edge of T (DEFINITION 6.32). Here we regard the stack as the vertex sequence of the corresponding path.

6.52 THEOREM.

Let G be any connected graph, and let v be a vertex of G. There exists a lineal spanning tree for G rooted at v.

Proof. The proof is by induction. If G has only one vertex the result is trivial. If G has more than one vertex, then, by connectivity, there is an edge of the form {v,v'}. This edge of G defines a lineal subtree T with two vertices. In general, let T be a lineal subtree of G with |V(T)| = k and assume k < n =

25

$|V(G)|$. We shall construct a lineal subtree T' of G with $k+1$ vertices. There is some vertex $w \in V(G)$ with $w \notin V(T)$. By connectivity, there is a path $w_1, w_2, \ldots, w_t$ ($w = w_1$) arriving at the set $V(T)$ for the first time at w_t. Thus, the edge $\{w_{t-1}, w_t\}$ has exactly one vertex (namely w_t) in common with $V(T)$ and, by the induction hypothesis, this vertex must be a vertex of STACK(e) where e is the preorder last edge of T. Let p be the preorder largest (i.e., furthest from the root r) vertex of STACK(e) for which there is a vertex $q \notin V(T)$ with $\{p,q\} \in E(G)$. Let T' be the tree formed by adding $\{p,q\} = e'$ to T. We show that T' is lineal with respect to G and the root v. By defnition, $\{p,q\} = e'$ is the preorder last edge of T'. We must show that T' satisfies the conditions of DEFINITION 6.51 (see FIGURE 6.53).

(1) Let $\{x,y\} \subseteq V(T')$ be an edge of G. We must show that x and y are a lineal descendant pair of T'. If $\{x,y\} \subseteq V(T)$ then the result is evident by the induction hypothesis. Otherwise, one of the vertices of this edge, say y, is equal to q. Again, by the induction hypothesis, we must have $x \in$ STACK(e). By the maximality of p, $x \in$ STACK(p). Thus x and $y = q$ are a lineal descendant pair.
(2) Suppose that $e \in E(G)$ has exactly one vertex x in common with $V(T')$. If $x = q$, then x is a vertex of STACK(e'). If $x \neq q$ then, by the induction hypothesis, x is a vertex of STACK(e). But again using the maximality of p, x is a vertex of STACK(p) and hence in STACK(e').

This completes the proof of the theorem since by induction, there is a lineal subtree of G with n vertices rooted at v. This subtree must be a lineal spanning tree rooted at v.

6.53 THE INDUCTION STEP OF THEOREM 6.52.

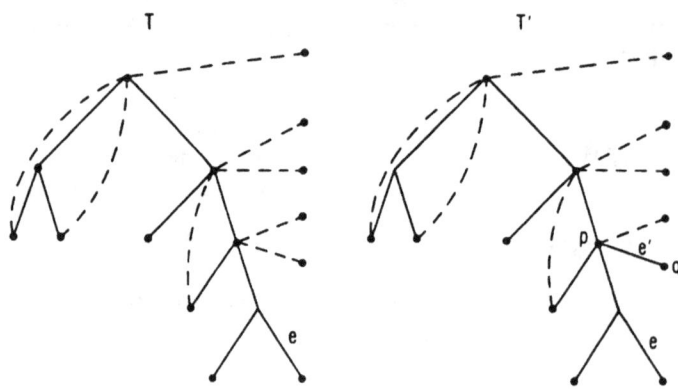

Solid lines are TREE edges, dotted lines edges of G.

Figure 6.53

We shall, in what follows, give many applications of the notion of a lineal spanning tree. An alternative way of viewing lineal spanning trees is suggested in EXERCISE 6.54.

6.54 EXERCISE.

Let $G = (V,E)$ be a connected graph, $v \in V$. Assume $|V| \geq 2$. Choose w adjacent to v. Let W denote all $x \in V$ for which there exists a path (not self-intersecting) from x to v with w as a vertex. Let $U = V - W$. By inductively constructing a lineal spanning tree for G restricted to W (root w) and G restricted to U (root v), prove THEOREM 6.52. *Hint:* See FIGURE 6.55.

6.55 IDEA FOR ALTERNATIVE PROOF OF THEOREM 6.52.

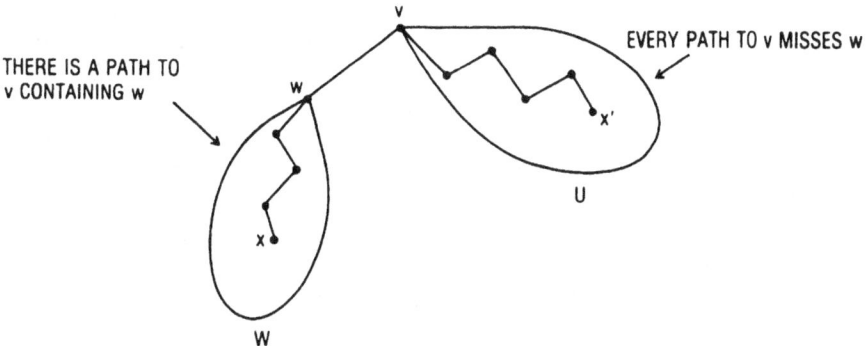

Figure 6.55

An algorithm for constructing a lineal spanning tree follows immediately from THEOREM 6.52 and follows exactly *procedure* 6.45. We need only describe the function SON for this particular case. The tree T_p will be the lineal spanning tree. Given v, the tree $T_1 = (\{v\}, \phi)$. The function $SON(x,T_i)$ is ϕ if all of the vertices adjacent to x in G are in T_i (i.e., $A_x - V(T_i)$ is empty where A_x is the set of vertices adjacent to x in G). If $A_x - V(T_i)$ is not empty, $SON(x,T_i)$ may be chosen to be any element of that set. As the choice of this element is left open one might say that the algorithm in this form is "undeterministic." Once a tree is constructed by this process, however, an order is defined on A_x for each vertex x of G (the order in which the vertices of A_x are examined to see whether or not they are vertices of the current tree). This ordering of the A_x makes G into an ordered graph. If, with respect to this order, we define $SON(x,T_i)$ to be the first element of $A_x - V(T_i)$ in each case, then we construct the same final tree. Henceforth, in constructing a lineal spanning tree for G we shall assume at the outset that the connected graph G is an ordered graph (without loops). In this case, $SON(x,T) = \phi$ if $A_x - V(T)$ is empty and otherwise equals the first element of $A_x - V(T)$ with respect to the order on A_x. Using the *procedure* 6.45 NEXT(T), we have

6.56 *procedure* **(construct a lineal spanning tree).**

Initially $T = (\{v\}, \phi)$;
while $|V(T)| < |V(G)| = n$ *do* $T := \text{NEXT}(T)$;

Procedure 6.56 is in fact a very efficient and easy algorithm to implement. In order to give a discussion of the "complexity" of this procedure it is necessary to have a simple model of computing. We next give such a model, intended only as a rough intuitive guide. The reader should review the discussion of basic data structures at the start of CHAPTER 1 (in particular, TABLE 1.14, FIGURE 1.15, 1.16, and 1.17, and related discussion). The discussion to follow is related to the "direct access model" as described in CHAPTER 1. We represent by a circle $\bigcirc$ a location where "elementary" data are to be stored (an integer, letter of the alphabet, pairs or triples of such symbols, etc.). Each circle has an address, written at the side of the circle when important to the discussion: $x \bigcirc$. Our basic assumption is that given the address of a circle, we can find it and read its contents in "constant time" (this is the definition of constant time for us!).

One obvious problem with these ideas is that if the name of a circle, x, is an integer and the problem required *very many* circles, then x might be a very large integer with 10^{10} digits. Just to scan the digits of such a large integer is going to require an enormous amount of time. We ignore this problem. The same problem arises in rigorous discussions of complexity (in the "random access model" of computation) and is ignored there also. We also do not make any attempt to define the class of possible symbols for naming circles or the possible contents of circles. We rely instead on "common sense" to dictate these choices. Our mathematical interests here do not lie in the direction of complexity analysis of algorithms. We wish only to gain some sensitivity to such problems in order to relate them to the combinatorics. We shall describe an algorithm and indicate how the computations are to be made. We then attribute costs to certain basic steps in the computation and call the total cost the measure of complexity of the algorithm. We hope to at least be reasonable enough to "get the ball rolling" in so far as complexity questions are concerned. We start with some examples.

If f is a function from a set $\underline{D}$ to a set R (assume D finite) then f can be represented in tabular form: $x \;\boxed{f(x)}$. Given x, we assume that we can find $f(x)$ in constant time. In particular, arrays such as a matrix (a_{ij}) can be represented in this form: (ij) $\boxed{a_{ij}}$. Thus, given i and j, we can read the ij^{th} entry of the matrix in constant time. One basic type of information that we shall store in our circles is a "pointer" to another circle: $x \;\textcircled{y}$ where y is the address of another circle, $y\bigcirc$. The symbol y is the *pointer* in location x. This situation is also represented by the diagram $x \bigcirc \rightarrow y \bigcirc$ or simply by $\bigcirc \rightarrow \bigcirc$ when the actual names x and y are not important to the discussion. Again, we suppose that the elements of D,R and the symbols that represent points (or addresses of circles) are, in some vague sense, "elementary." If we wished to list the three symbols # $ ¢ in that order we might do so by using pointers as follows: start at x, x(#y), y($z) z(¢). In location x we read # and go to y were we read $ and go to z, etc. We must, of course, have some conventions for telling pointers from data inside a circle, but this is clear from the notation in this example. Once the pointers are given, the order or location of the corresponding circles in the text obviously has no importance as we have assumed that, given the name, a circle can be located in constant time. Thus the same list could be given by the following: start at x, y($z), z(¢,) x(#y). Such a method for listing a sequence of symbols is called a "linked list" (see FIGURE 1.15, CHAPTER 1). It is an example of what we shall refer to informally as a "data structure." The linked list structure can be indicated by the following notation: $\bigcirc \rightarrow \bigcirc \rightarrow$ ... $\rightarrow \bigcirc$ (see FIGURE 1.16, CHAPTER 1).

The amount of work involved in an algorithm will depend very much on the data structure used. If one specifies an ordering of a sequence of symbols by listing them one after the other in the text (for example, #$¢) then to insert a new symbol, say &, between two others, say # and $, requires shifting all symbols to the right of the new symbol added by one space (for example #$¢ becomes #&$¢). Thus the amount of work involved in inserting a new symbol depends on the length of the list. In the case of the linked list the amount of work will not depend on the length of the list. (See the discussion associated with FIGURE 1.15, CHAPTER 1). For example, to insert & between # and $ we modify the above linked list to obtain the following: start at x, y($z), z(¢) x(#w) w(&y). (In general, we do not redraw the old circles but simply add the new circle and change the relevant pointers. If the old circles were redrawn then the changes would clearly not be independent of the length of the list! This should be thought of in terms of FIGURE 1.15 of CHAPTER 1.) The pointer in circle x was changed from y to w and a new circle named w was added with the symbol & and a pointer y. These changes do not depend on the length of the list.

One problem here that may or may not be minor is that of locating the name of the circle containing the symbol after which the new symbol is to be inserted. If we are asked to "insert & just after the symbol contained in x" then this is no problem by our assumptions. If, on the other hand, we are asked to "insert & just after $" then we must first find the name of the circle containing $. One possibility is to keep a *function* or *array* $\begin{pmatrix} \# & \$ & ¢ \\ x & y & z \end{pmatrix}$ that gives the location of each symbol. This means that x, y, and z become contents of circles, and #, $, and ¢ (or perhaps numerical codes associated with these symbols if that is preferred for some reason) become addresses or names of circles. Now when asked to "insert & just after $" one obtains the location of $ from the array and modifies the linked list accordingly. One then must modify the array to give the location of &. These modifications still require constant time, independent of the length of the list in our model.

We now use the above ideas to discuss the complexity of the problem of finding a lineal spanning tree of a graph. Consider the graph of FIGURE 6.57.

6.57 A GRAPH G.

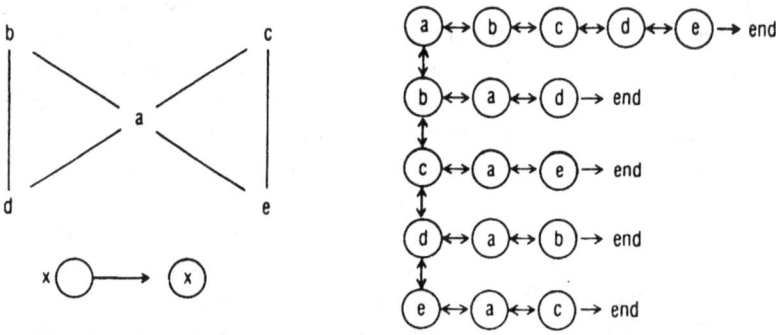

Figure 6.57

The "adjacency table" of FIGURE 6.57 has vertices as contents of circles with pointers as shown. A list with pointers both to the next element of the list and the previous element of the list is called a "doubly linked list." Thus, the list of vertices (the vertical list of FIGURE 6.57) is a doubly linked list. Associated with each vertex x there is a pointer $x\bigcirc \rightarrow \circledx$ to the location of x. The lists of adjacent elements are also doubly linked. We will not need all of these pointers for the lineal spanning tree problem. In fact, the doubly-linked list structure of the list of vertices will not be used at all for the lineal spanning tree problem. As this data structure is useful for other problems, it is a good general convention for the list of vertices. As remarked, we assume that given a vertex name x, we can go in constant time to the location where x is stored in the list of vertices. How this can be accomplished within the framework of our model was explained in the previous paragraph and in FIGURE 1.18 of CHAPTER 1. We shall construct a lineal spanning tree for G of FIGURE 6.57 rooted at a. G is an ordered graph as given by the adjacency table of FIGURE 6.57. We start with T = ({a},ϕ) and construct a sequence of trees using NEXT(T), *procedure* 6.45, until a lineal spanning tree is obtained (using *procedure* 6.56). The lineal subtree (DEFINITION 6.52) obtained at each stage of the algorithm will be denoted by T. For each such tree, we shall have a function VERT: V(T) $\rightarrow$ {0,1} defined by VERT(x) = 1 if x is a vertex of T and 0 otherwise. Initially S denotes the stack of the preorder last element of T. *We regard T as directed in the natural fashion* (away from the root, DEFINITION 6.25).

We now carry out the construction of a spanning tree step by step, showing at each stage the basic data structures: G, T, S, and VERT.

30

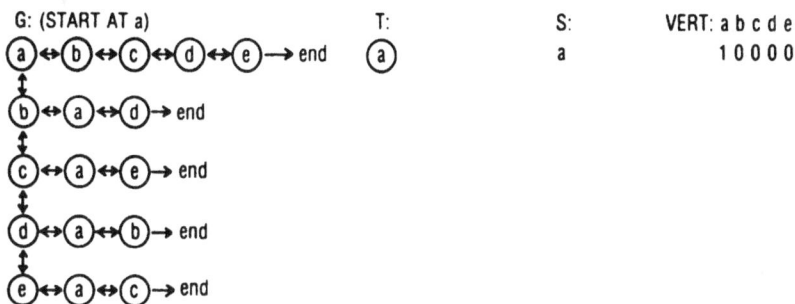

Start at a and check the first vertex b adjacent to a (follow pointer to right). VERT(b) is 0 so b must be added to T and the stack S and deleted from G. Then VERT(b): = 1. We then go to b (vertical list).

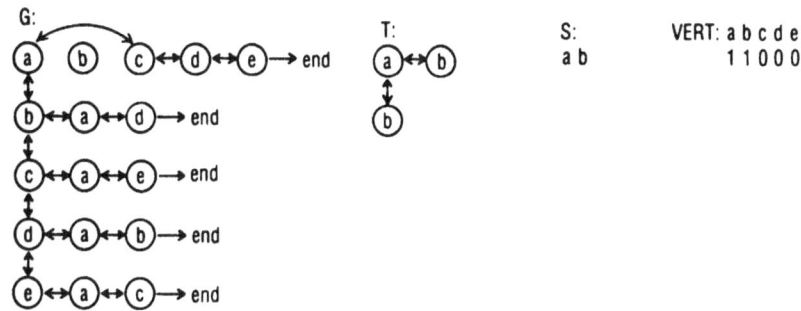

Start at b and check the first vertex in its adjacency list, a. VERT(a) = 1 so a has already been added to the tree. Change the pointers to d:

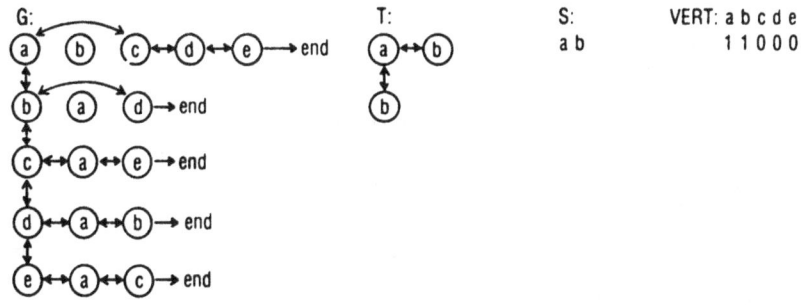

The pointer from b now points to d. VERT(d) = 0 so d has not been added to the tree T. Change the pointer from b in G to indicate that all vertices in its adjacency list have been considered, add d to T, add d to S, and set VERT(d) = 1.

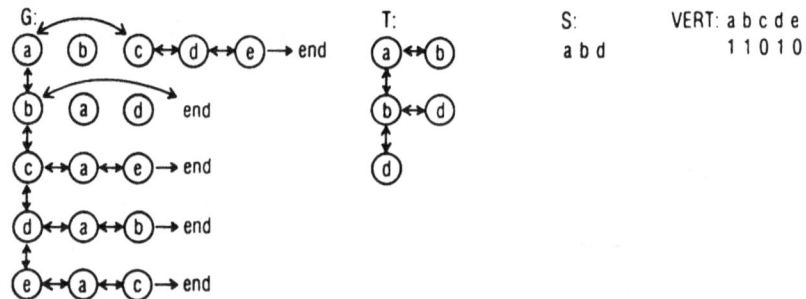

The first pointer at d points to *a*. VERT(a) = 1. Change the pointer to b. VERT(b) = 1. Change the pointer to "end" and remove d from S. Now b is the last element of S:

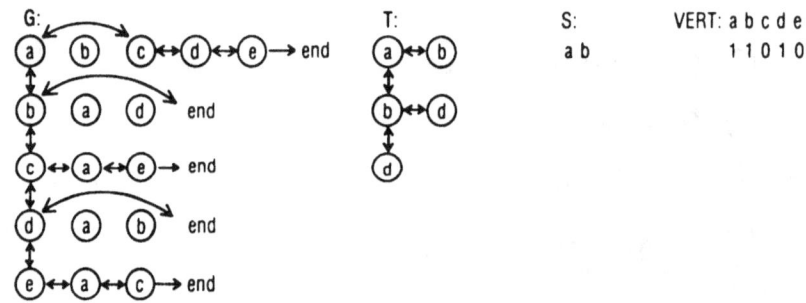

Start at b. The pointer points to "end" so remove b from S to get S = *a*. Go to *a* in G and follow pointer to c. VERT(c) = 0 so add c to S, change pointer to d, add c to S and T, change VERT(c) to 1 and go to c.

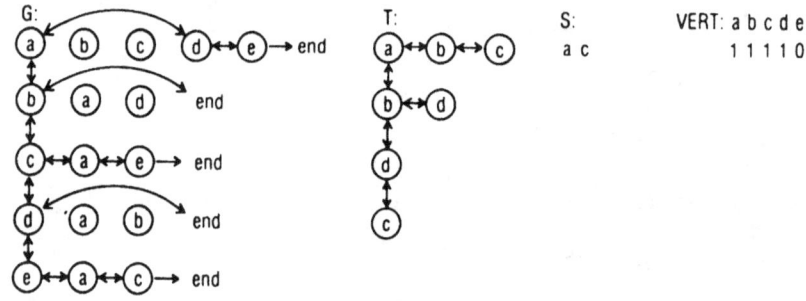

Start at c. VERT (a) = 1. Change pointers to e. VERT(e) = 0 so add e to S, VERT, T:

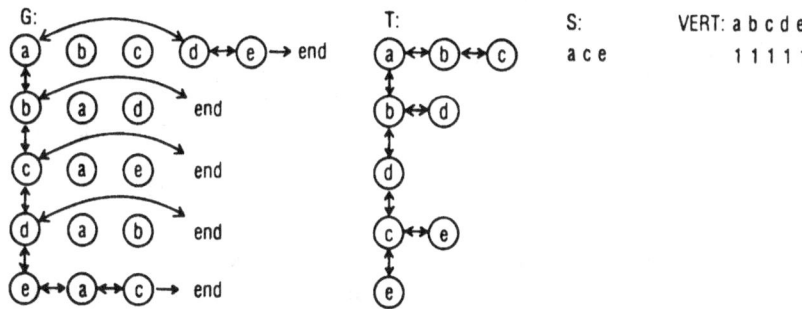

Start at e. The first pointer is to *a*, but VERT(a) = 1. Change the pointer to c. VERT(c) = 1. Change pointer to end and remove e from S. Last element of S is now c so go to c. The pointer at c points to end so remove c from S and go to last element of S which is now *a*. The pointer at *a* points to d but VERT(d) = 1 so change pointer to e. VERT(e) = 1 so change pointer to end and remove *a* from S. S is now empty so stop. The terminal configuration is shown in FIGURE 6.58.

6.58   TERMINAL DATA STRUCTURE AND LINEAL SPANNING TREE.

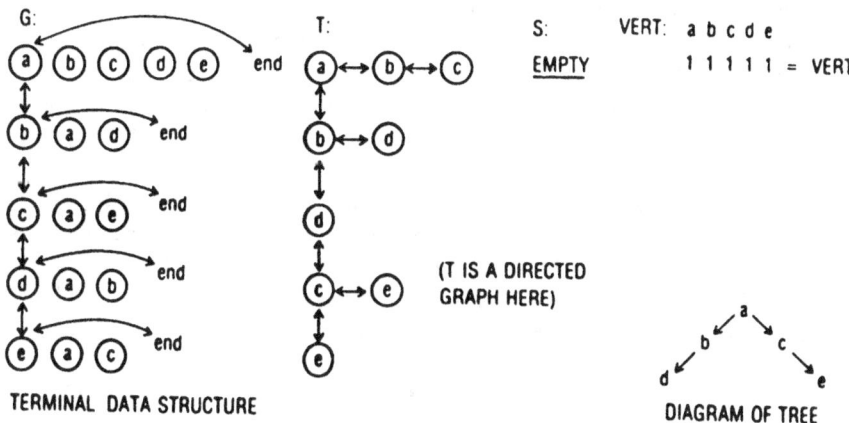

TERMINAL DATA STRUCTURE   DIAGRAM OF TREE

Figure 6.58

We assert that the time required to construct the spanning tree of a connected graph using the above data structures is proportional to the number of edges in the graph. As the algorithm is carried out, certain basic operations are performed (going to a specified address, reading the last element of S, changing a pointer, etc.). We observe that these basic operations can be partitioned among the edges

33

of the graph in a natural way such that the number of basic tasks assigned to each edge is a constant independent of the number of edges in G. The first thing we do in each case is read the last entry of S. For example, if S = a b c, we read c. We then go to this vertex in the list of vertices of G. If the vertex x is read from S, and y is the vertex obtained by following the pointer from x into its adjacency list, then, depending on the value of VERT(y), certain pointers must be changed—S might have y added to it, and VERT might have to be updated. Clearly, these operations, for fixed x and y do not depend on $|E|$. We attribute these costs to the edge {x,y}. The reader can easily verify that all costs can be partitioned among the edges in this fashion. The maximum cost that we would attribute to any edge in this way is a constant indpendent of $|E|$. Thus we say that the complexity of this algorithm is bounded above by a constant times the number of edges. Clearly, all edges of a connected graph must, in general, be examined to determine a lineal spanning tree in this fashion. Thus we say that the complexity of this algorithm is linear in the number of edges of G (i.e., proportional to a constant times the number of edges). In fact, of course, we have an inequality $c_1|E| \leq COST \leq c_2|E|$ for constants c_1, c_2. In the adjacency structure for G (see FIGURE 6.57) the first column represents the linked list of vertices for G. Notice that we go to a vertex in this list as many times as that vertex appears as the last element of S in the course of the algorithm. This is equal to the degree of that vertex in the lineal spanning tree and in general is not a constant independent of $|E|$. We assert only that *for each x in the vertex list and each y adjacent to x*, the cost of processing that y is (as explained above) bounded above by a constant. Finally, we remark again that a statement such as "go to x" where x is a vertex does not refer to the name or "address" of a circle but rather to its contents. In order to be able to go to an address in constant time given the symbol representing its contents, an array representing this correspondence is kept (as discussed above, and shown in FIGURE 1.18 of CHAPTER 1).

Note that the data structure used to represent the spanning tree T in FIGURE 6.58 is the basic adjacency list valid for all graphs. Knowing that T is a tree, a better data structure would be the data structure used in FIGURE 1.87 of CHAPTER I. This data structure is used there for binary trees but extends in an obvious way to arbitrary trees (each node still has a pointer to its father, its next brother, and its first son).

We now consider an application of the lineal spanning tree algorithm to the generation of connected graphs.

6.59 DEFINITION.

Let G be a connected graph and T a spanning tree of G. An edge of G that is not an edge of T is called a *chord*. If T is a lineal spanning tree rooted at v then a chord is called a *backedge* and the *natural direction* of a backedge {x,y} is (y,x) where y is a descendant of x in T.

For example, the graph of FIGURE 6.57 has the lineal spanning tree shown in FIGURE 6.58. There are two directed chords, (d,a) and (e,a). Both are backedges since T is lineal.

We refer now to *procedure* 6.56 and the discussion (just preceding *procedure* 6.56) of the function SON(x,T). Suppose that the vertex set V of G is a set of nonnegative integers. Order each adjacency list by decreasing order on integers. Thus SON(x,T) always picks the largest vertex (as an integer) from among the vertices adjacent to x in G but not already added to T. Let T denote the lineal spanning tree that results from *procedure* 6.56 with this choice of SON. Suppose that (y,x) is a directed backedge of G with respect to T. There is a unique path x,t,. . .,y in T from x to y and the length of this path must be at least 2. We claim that t > y as integers. Consider the point in *procedure* 6.56 when t was the preorder last element in the tree T (a subtree of the final spanning tree). Note that at this point y was not a vertex of the current tree. Let T′ denote the tree just prior to adding t. Evidently, SON(x,T′) = t and not y. But, both t and y are adjacent to x in G and both are not vertices of T′. By the maximality property of SON we must then conclude that t > y.

6.60 DEFINITION.

Let T = (V,E) be a rooted tree with V linearly ordered. Let *a* be a vertex of T. A pair of vertices of T, (a,b), with b a descendant of *a* and b < *a* is called an inversion of T. We let I(T) denote the set of all inversions of T.

6.61 THE SET OF INVERSIONS.

I(T) = {(5,3),(6,4)}, V = {0,1,. . .,6}:

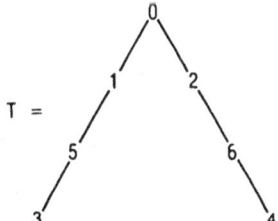

In light of DEFINITION 6.60 and the discussion just preceding, we have THEOREM 6.62. We denote by MAXSON the version of SON previously discussed that always selects the maximum available son.

6.62 THEOREM.

Let G be a connected graph and let T be a lineal spanning tree constructed using MAXSON. If (y,x) is a directed backedge of G relative to T, and x,t,. . .,y is the path from x to y in T, then (t,y) is an inversion.

Let CON(V) denote the set of connected graphs on a set of non-negative integers V and let TR(V,v) denote the set of trees on V rooted at v where v is the minimum of V. Using the procedure MAXSON we have shown how, given a connected graph G, to construct a lineal spanning tree rooted at v. Call this tree T. Define a subset S of the inversions I(T) by (t,y) ∈ S if (y,x) is a backedge of G relative to T where x,t,. . .,y is as in 6.62. Denote the mapping INV by the rule INV(G) = (T,S).

6.63 EXERCISE.

Show that the mapping INV is a bijection between the set CON(V) and the set $\{(T,S): T \in TR(V,v)$ and $S \subseteq I(T)\}$.

We now associate with each integer n a polynomial that we call the *inversion enumerator*. Let $V = \{0,1,. . .,n\}$ and consider the sum $\Sigma t^{|I(T)|}$ where the sum is over all T in TR(V,0). This sum can be written in the form $p_n(t) = \sum_{s=0}^{\binom{n}{2}} a_s^{(n)} t^s$. This polynomial is the *inversion enumerator polynomial*. It is evident that $a_s^{(n)} = |\{T : |I(T)| = s\}|$. The first four such polynomials are $p_1(t) = 1$, $p_2(t) = t + 2$, $p_3(t) = t^3 + 3t^2 + 6t + 6$, $p_4(t) = t^6 + 4t^5 + 10t^4 + 20t^3 + 30t^2 + 36t + 24$. In the following exercise we develop some of the remarkable properties of these polynomials. (Note: $p_0(t) = 1$.)

6.64 EXERCISE.

(1) Show that the polynomials $p_n(t)$ satisfy the recursion

$$p_{n+1}(t) = \sum_{i=0}^{n} \binom{n}{i} p_i(t) p_{n-i}(t)(1 + t + t^2 + \ldots + t^i).$$

Hint: This reflects a basic recursive way of constructing TR(V,0) where $V = \{0,1,. . .,n+1\}$. Consider first the principal subtree of the root, say T_1, that contains the vertex n + 1. The number of vertices of this subtree can be 1,2,. . .,n+1. Thus we must choose, in each case, i other vertices besides the vertex n + 1, i = 0,1,. . .,n $\left(\binom{n}{i}\text{ choices}\right)$. For a fixed choice we must then choose a root. Finally, for a fixed choice we must consider the tree $T - T_1$ gotten by deleting T_1 from T.

(2) Show that $p_n(0) = n!$ by constructing a bijection between all trees T ∈ TR(V,0) with I(T) empty (no inversions) and all permutations of $\{1,. . .,n\}$ = $\underline{n}$. It is evident that $p_n(1) = (n+1)^{n-1}$ which is the number of trees on V. Do we need to know in advance that $(m+1)^{m-1}$ counts the number of trees, or does this follow from problem (1) above? Also, from the definition of p_n it is immediate that $p_n(2)$ is the number of connected graphs. Show

36

that the number of such directed graphs on V is given by $2^n p_n(3)$. A permutation f of $\underline{n}$ is alternating if $f(1) < f(2)$, $f(2) > f(3)$, $f(3) < f(4)$, etc. For example, if n = 3, there are two alternating permutations 1 3 2 and 2 3 1. For n = 4 there are five alternating permutations 1 3 2 4, 1 4 2 3, 2 4 1 3, 2 3 1 4, 3 4 1 2. In general, let E_n (the so called Euler number), denote the number of alternating permutations of $\underline{n}$. It is a remarkable fact that $p_n(-1) = E_n$ (try and prove it or see the references).

(3) The recursion of EXERCISE (1) contains a method for actually constructing trees classified by number of inversions. Using this idea, or any other you might think of, discuss the problem of generating all trees in TR(V,0) with a fixed number, say k, of inversions.

THEOREM 6.62 and EXERCISE 6.63 provide an interesting algorithm for constructing all connected graphs on V = {0,. . .,n}. One first constructs all trees on V, rooted at 0. For each such tree T one then constructs all subsets S of I(T). For each such set S, one includes exactly those backedges corresponding to elements of S. The tree T can be constructed by any method (for example using INVPRU, *procedure* 6.21). We illustrate the idea by constructing all connected graphs on {0,1,2,3}.

6.65 CONNECTED GRAPHS WITH FOUR VERTICES.

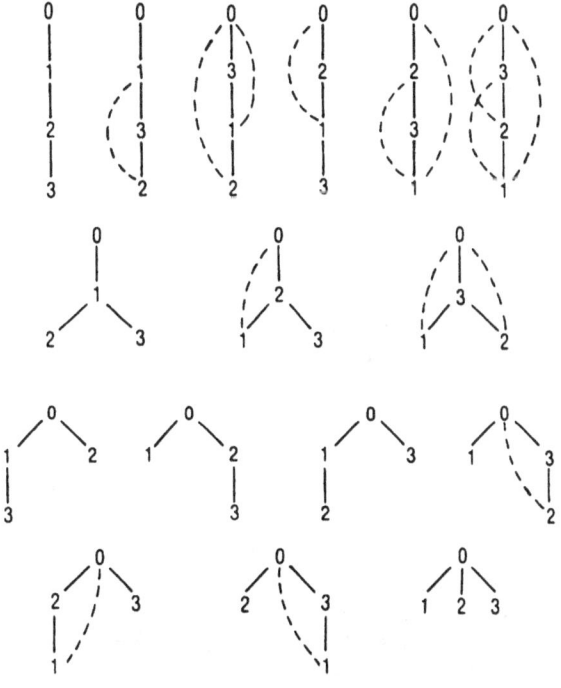

Figure 6.65

37

The dotted lines refer to edges that may be included or excluded. By including and excluding these edges in all possible ways, all connected graphs with vertex set V = {0,1,2,3} are obtained.

6.66 STRUCTURE OF AN ORDERLY ALGORITHM.
$B(R_{i+1}) \subseteq R_i$ IF AND ONLY IF $R_{i+1} \subseteq B^{-1}(R_i)$.

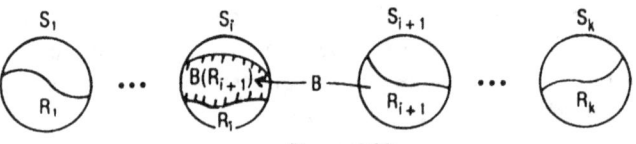

Figure 6.66

In FIGURE 6.65 it was necessary to construct all trees on a given vertex set. One method, as suggested there, is to use the map INVPRU (*procedure 6.21*). As an interesting aside, we consider another quite different method that is very general and has a wide range of applications. This method was discussed in CHAPTER 4 (see FIGURE 4.65). Let S be a set and R a subset of S. Suppose that the elements of S are to be examined one by one and a test is to be performed to see whether or not an element is in R. In practice, there are often ways to avoid examining all of the elements of S. The method we now give formalizes one such method. Suppose that S is partitioned into k disjoint subsets (called blocks) $S_1, \ldots, S_k$ and let R_i denote $S_i \cap R$. For i = 1, \ldots, k-1, let B be a function mapping S'_{i+1} to S_i where S'_{i+1} contains R_{i+1} and is contained in S_{i+1}. We assume that $B(R_{i+1})$ is a subset of R_i for all i = 1, \ldots, k-1. (See FIGURE 6.66.) The following defines an "orderly algorithm for the function B."

6.67 ORDERLY ALGORITHM.

Start with R_1 given. For each i = 1, \ldots, k-1, examine the elements of $B^{-1}(R_i)$ and test to find all elements in R_{i+1}.

Observe that the condition $B(R_{i+1})$ is contained in R_i implies that the orderly algorithm examines all elements of R. The case k = 1 covers the basic search procedure, so in a sense, all such algorithms are "orderly algorithms" for the trivial B. The interest and challenge is to choose a good partition and function B to greatly reduce the number of elements of S *not in* R that are examined. Also, the function B must be easy to compute in some sense. For example, one could trivially define B(x) = y for all x in R_{i+1} where y is some fixed element in R_i. Such a choice of B would define an orderly algorithm but would be of little practical use in general. These ideas are best illustrated by giving an example. We consider now the example of generating trees. It is helpful at this point to consider the notion of isomorphism of graphs.

For any set S, let $\mathscr{P}_2(S)$ denote the set of all subsets of size 2 of S. Let V and V' be sets and let f denote a bijection between V and V'. Then f induces a bijection $\underline{f}$ between $\mathscr{P}_2(V)$ and $\mathscr{P}_2(V')$ by the rule $\underline{f}\{x,y\} = \{f(x),f(y)\}$. For any graph G, define $\underline{f}(G) = (f(V),\underline{f}(E))$ where $G = (V,E)$ and f is a bijection between V and V'. Thus $\underline{f}(G)$ is a graph with vertex set $V' = f(V)$.

6.68 DEFINITION.

Two graphs $G = (V,E)$ and $G' = (V',E')$ are *isomorphic* if there is a bijection f between V and V' such that $G' = \underline{f}(G)$.

6.69 EXERCISE.

Let BIJ(A,B) denote the set of bijections from A to B. Define a map F from BIJ(V,V') to BIJ($\mathscr{P}_2(V),\mathscr{P}_2(V')$) by $F(f) = \underline{f}$. Prove carefully that $\underline{f}$ is in fact a bijection on $\mathscr{P}_2(V)$ to $\mathscr{P}_2(V')$ and show that F is an injection if $|V| > 3$.

If A is a set then we use the notation PER(A) to denote the set of all per-mutations of A. PER(A) is a group under compositions of permutations in the usual manner (the "symmetric group on A") (PER($\underline{n}$) = S_n). If K is any subgroup of PER(A), then we may define an equivalence relation on A by $x_K y$ if there exists a permutation k in K such that $k(x) = y$. This relation is easily seen to be reflexive, symmetric, and transitive. The equivalence classes of this relation are called the *orbits* of K in A. The reader interested in thinking more about these ideas should read the beginning of CHAPTER 4, TOPIC III (DEFINITION 4.1 to LEMMA 4.9).

Let $\mathscr{P}(A)$ denote the set of all subsets of A (the "power set of A"). If f is an element of PER(A) then f also permutes the elements of $\mathscr{P}(A)$ in a natural way: for S in P(A), define $f(S) = \{f(x): x \text{ in } S\}$. We call this permutation group on $\mathscr{P}(A)$ the *subset action* of PER(A). Consider the situation of EXERCISE 6.69 with $V = V'$. In this case, F is an injection between PER(V) and PER($\mathscr{P}_2(V)$). It is easily seen that for f and g in PER(V), $F(fg) = F(f)F(g)$, and thus F is an *isomorphism* of PER(V) to a subgroup of PER($\mathscr{P}_2(V)$). We call this subgroup the *edge action* of PER(V) and denote it by EPER(V). Clearly, EPER(V) is the subset action of PER(V) restricted to $\mathscr{P}_2$, the subsets of size 2. We may consider also the subset action of EPER(V). Thus each element $\underline{f}$ of EPER(V) acts on subsets E of $\mathscr{P}_2(V)$. Each such subset E defines a graph $G = (V,E)$ of GRAPHS(V), the set of all graphs on V, and hence the subset action of EPER(V) defines a permutation group on GRAPHS(V). Thus we have four basic permutation groups: PER(V), EPER(V), the subset action of EPER(V), and the action of EPER(V) on GRAPHS(V). All four are isomorphic as groups to PER(V). As an example, consider $V = \{1,2,3\}$. Let $E = \{\{1,2\},\{1,3\}\}$. The group PER(V) has six ele-ments. One such element is $\begin{pmatrix} 1 & 2 & 3 \\ 2 & 3 & 1 \end{pmatrix} = f$. Thus $\underline{f}(E) = \{\{2,3\},\{2,1\}\}$. Another element is $\begin{pmatrix} 1 & 2 & 3 \\ 1 & 3 & 2 \end{pmatrix} = g$. We have $\underline{g}(E) = \{\{1,3\},\{1,2\}\} = E$. The action of $\underline{f}$

and $\underline{g}$ on the corresponding graphs is shown in FIGURE 6.70 (the notation $\underline{f}$ and $\underline{g}$ of DEFINITION 6.68 is dropped here).

6.70 PERMUTATIONS OF V ACTING ON G = (V,E).

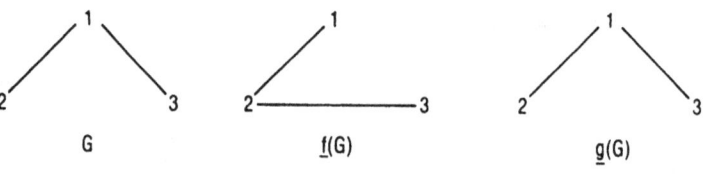

Figure 6.70

Observe that the permutation g applied to the graph G in FIGURE 6.70 does not change G. Thus G is a fixed point of g. Such a g is called an *automorphism* of G or a *stabilizer* of G. The set of all elements of PER(V) that are automorphisms of a given graph G are easily seen to form a group called the *automorphism group* of G or the *stability subgroup* of G. Let AUT(G) denote this subgroup of PER(V). Clearly, two graphs of GRAPHS(V) are isomorphic *if and only if they are in the same orbit of* EPER(V) *acting on* GRAPHS(V). Thus the number of distinct isomorphism classes of graphs is the number of orbits of this action. Given G, the number of graphs isomorphic to G (the number of elements in its orbit) is, by an elementary result from group theory, (LEMMA 4.9, Chapter 4), the ratio n!/|AUT(G)|. Thus, there are three graphs isomorphic to G of FIGURE 6.70. Conceptually, one may generate all graphs in the orbit of a given graph G by drawing the diagram of G, erasing all vertex labels leaving only dots, labeling the dots with all possible permutations of the original vertex labels, and then throwing out all duplicate copies of graphs. This process is illustrated for the graphs of FIGURE 6.70 in FIGURE 6.71. The pattern of dots (called the "unlabeled graph" in some books) is shown in FIGURE 6.71(b). FIGURE 6.71(a,c,d,e,f,g) give the six possible ways of labeling FIGURE 6.71(b). Note that (a) is the same as (c), (d) is the same as (e), and (f) is the same as (g) *as graphs*. Thus, there are exactly three graphs in the isomorphism class associated with the dot diagram (b), namely, (a), (d), and (f) as indicated in FIGURE 6.71.

6.71 KEEP a,d,f.
AN ORBIT CLASS OF THE GRAPH OF FIGURE 6.70.

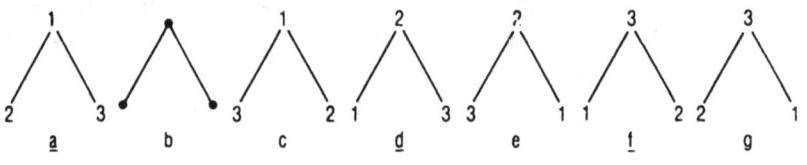

Figure 6.71

The diagram for a graph G such as shown in Figure 6.71(b) is a convenient symbolic way of representing the orbit or isomorphism class of G. We shall call such a diagram an *orbit diagram* for G. Such diagrams, as mentioned above, are often called "unlabeled graphs." We prefer not to use that terminology here, since "labeled graphs" will, for us, be graphs plus something extra (called labels). When these labels are removed we again have graphs as we have defined them, not orbit classes as above.

Orbit classes for ordered graphs may be defined in exactly the same manner. Consider, for example, the graph G of EXAMPLE 6.8 with vertex set {a,b,c,d} = V. Let $\begin{pmatrix} a & b & c & d \\ b & c & d & a \end{pmatrix}$ be in PER(V). This permutation is applied both to the edge set and to the adjacency table of G to obtain the resulting ordered graph as shown in FIGURE 6.72.

6.72 THE PERMUTATION $f = \begin{pmatrix} a & b & c & d \\ b & c & d & a \end{pmatrix}$ APPLIED TO THE EXAMPLE 6.8.

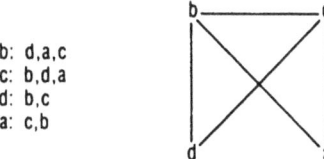

b: d,a,c
c: b,d,a
d: b,c
a: c,b

Figure 6.72

The orbit class of the ordered graph of FIGURE 6.72 can be specified by a diagram such as FIGURE 6.73. Order is indicated on incident edges by numerical labels.

6.73 THE ORBIT CLASS OF THE ORDERED GRAPH FIGURE 6.72.

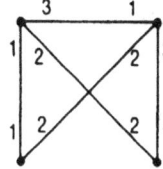

Note: The numbers closest to a given vertex specify the order of the edges incident on that vertex.

Figure 6.73

Orbit classes of rooted trees have diagrams such as that of FIGURE 6.74.

6.74 THE ORBIT CLASS OF A ROOTED TREE (ROOT CIRCLED).

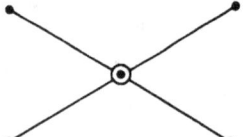

This type of structure is sometimes called a "rooted planar tree."

Figure 6.74

In the case of ordered rooted trees we draw the diagram in the standard manner with the root at the top and order specified by left to right orientation of edges as in FIGURE 6.75. In this manner the numerical labels used for ordered graphs in the general case are not required (see FIGURE 6.73) and the root is not circled.

6.75 THE ORBIT CLASS OF AN ORDERED ROOTED TREE.

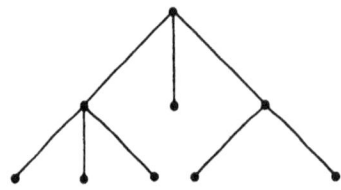

Figure 6.75

We now return to the discussion of "orderly algorithms." Referring to the ORDERLY ALGORITHM 6.67 and the discussion just preceding it, we take S to be the set of orbit diagrams of ordered rooted trees. The subset R will be a set of "canonical" diagrams. R will correspond in a natural way to the set of orbit diagrams of rooted trees (not ordered). The following definition is now necessary.

6.76 DEFINITION.

Let $\underline{T}$ be the orbit diagram of an *ordered* rooted tree T. The CODE($\underline{T}$) is the sequence obtained from the depth first sequence of edges of T, DFE(T) (see FIGURE 6.27 and DEFINITION 6.30) by replacing each first occurrence of an edge by 0 and each second occurrence by 1.

For example, the code of FIGURE 6.75 is 0010101101001011. This sequence is easily obtained geometrically by traversing FIGURE 6.75 in the same manner as indicated by the arrows of FIGURE 6.27(b). Each straight down arrow is replaced by a 0 and each straight up arrow is replaced by 1. When there is no danger of confusion, we shall write simply "tree" rather than "diagram of orbit class of a tree."

6.77 DEFINITION.

Canonical diagrams for ordered rooted trees: The orbit diagram $\underline{T}$ of an ordered rooted tree is canonical if it has one vertex or if

(1) Each principal subtree is canonical.
(2) The principal subtrees are arranged in nonincreasing order by size (number of vertices) from left to right.
(3) If two principal subtrees have the same size then the code (DEFINITION 6.76) of the one on the left is lexicographically less than or equal to the code of the one on the right.

There is a natural bijection between the *canonical* orbit diagrams of *ordered* rooted trees and orbit diagrams of rooted (not ordered) trees. The idea here is that when order does not matter for orbit diagrams of rooted trees we might as well always draw these diagrams such that they are canonical in the sense of DEFINITION 6.77. This drawing is unique and gives the desired bijection.

We now give an orderly algorithm for generating the canonical orbit diagrams R in the set of all orbit diagrams S of ordered rooted trees.

6.78 CANONICAL DIAGRAMS IN THE SENSE OF DEFINITION 6.77.
(a) IS CANONICAL   (b) IS NOT.

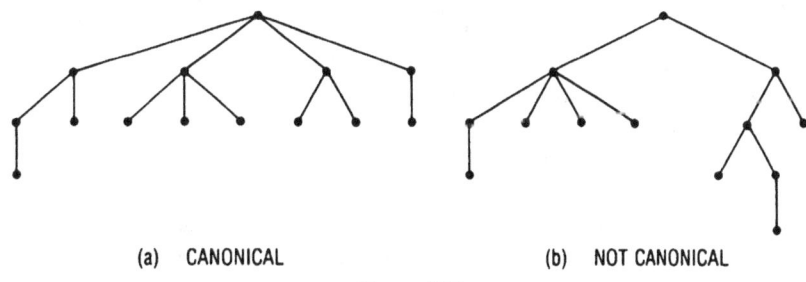

(a)   CANONICAL   (b)   NOT CANONICAL

Figure 6.78

We refer now to the ORDERLY ALGORITHM 6.67 and the preceding discussion. We classify the elements of S according to the number of edges i. The set S_i will be all orbit diagrams with i edges. S_1 contains only one element and it is canonical ($R_1 = S_1$). We now must define the mapping B (of the ORDERLY ALGORITHM 6.67) from S'_{i+1} to S_i such that $B(R_{i+1}) \subseteq R_i$. As in most examples of orderly algorithms, we shall take $S'_i = S_i$. If $\underline{T}$ is in S_i, then $B(\underline{T})$ is the orbit diagram obtained by removing the preorder last edge from $\underline{T}$ (see DEFINITION 6.31). An example is shown in FIGURE 6.79. The pendant vertex is also removed.

43

6.79 THE BASIC MAPPING B FOR THE ORDERLY ALGORITHM.

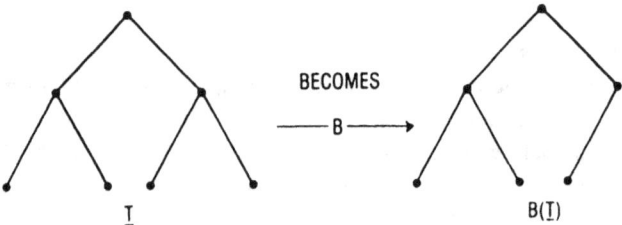

Figure 6.79

6.80 LEMMA.

If $\underline{T}$ is in R_{i+1} then $B(\underline{T})$ is in R_i.

Proof. The proof is a trivial induction on the number of edges in $\underline{T}$. If there are just two edges the result is obvious. Otherwise, consider a $\underline{T}$ with p edges. Suppose the lemma is true for all $\underline{T}$ with fewer than p edges. Let $\underline{T}'$ denote the rightmost principal subtree of the root of $\underline{T}$. If $\underline{T}'$ has one edge then it is the preorder last edge of $\underline{T}$. In this case, it is immediate from the definition of R_i that $B(\underline{T})$ is in R_i. Otherwise, $B(\underline{T}')$ is the rightmost principal subtree of $B(\underline{T})$. By the induction hypothesis, $B(\underline{T}')$ is canonical (in R_j for some j). All other principal subtrees of $B(\underline{T})$ are canonical (DEFINITION 6.77(1)). As $\underline{T}$ satisfies (2) of DEFINITION 6.77, it is immediate that $B(\underline{T})$ does also. Condition (3) is also immediate if one observes that the size of the rightmost principal subtree of $B(\underline{T})$ is now *strictly* less than that of any other principal subtree. This completes the proof.

It is evident from the proof of LEMMA 6.80 that in fact $B(R_{i+1}) = R_i$. The ORDERLY ALGORITHM 6.67 requires that $B^{-1}(R_i)$ be computed, i = 1,2,...,k − 1. For that we use the following rule for computing B^{-1}, described in 6.81.

6.81 RULE FOR COMPUTING B⁻¹.

Let $\underline{T}$ be in R_i. To compute $B^{-1}(\underline{T})$, construct one tree for each vertex of the stack of the preorder last vertex of $\underline{T}$ by adding an edge to that vertex.

An example of computing B^{-1} is given in FIGURE 6.82.

6.82 COMPUTING B^{-1}.
IN THIS EXAMPLE $\underline{T}$ IS NOT CANONICAL.

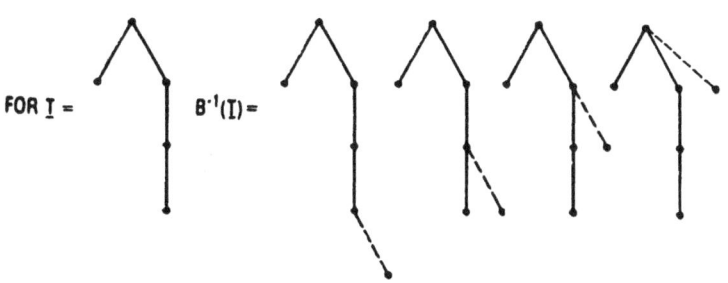

Figure 6.82

We now use the orderly algorithm to construct the first few R$_i$

6.83 CONSTRUCTION OF THE FIRST 4 R$_i$.
NC = NOT CANONICAL.

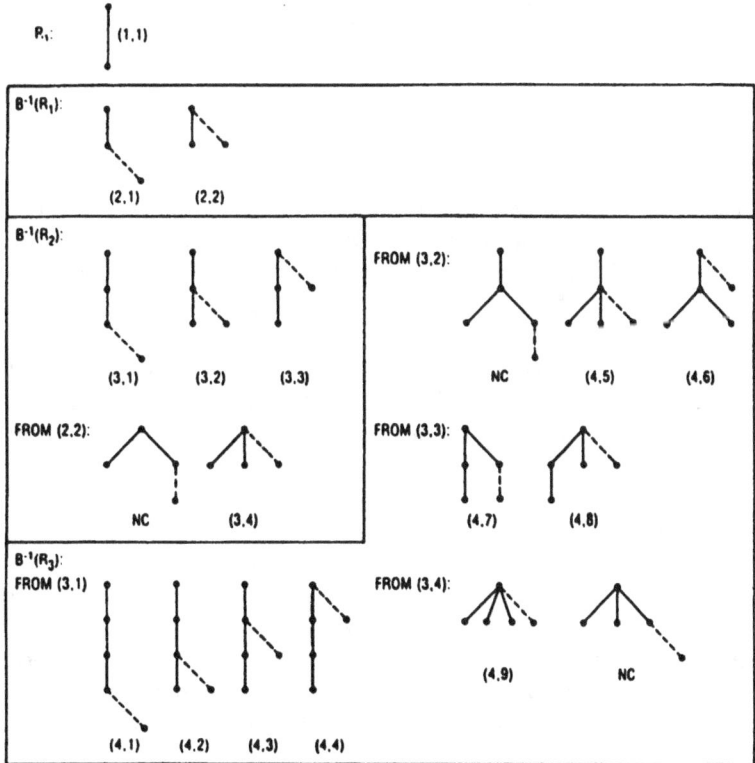

Figure 6.83

There is a natural tree structure on the trees (i,j) that are constructed by the orderly algorithm as in FIGURE 6.83. In this tree the root is (1,1). A tree (i′,j′) is a son of (i,j) if it is obtained from (i,j) by adding one edge. Only canonical trees are vertices. See FIGURE 6.84.

6.84 TREE STRUCTURE OF THE ORDERLY ALGORITHM OF FIGURE 6.83.

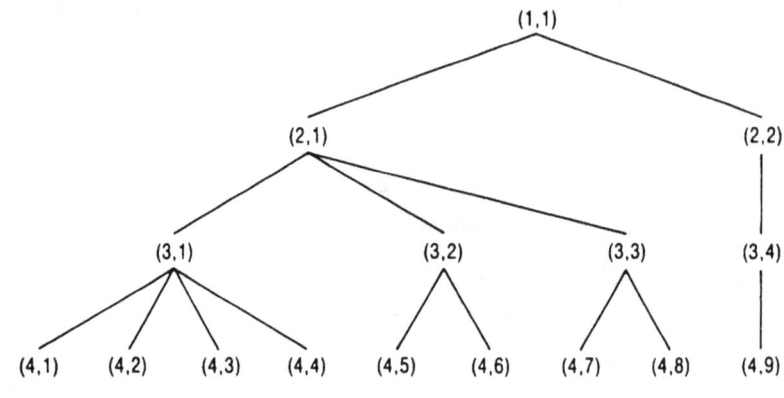

Figure 6.84

6.85 EXERCISE.

In FIGURE 6.83 only three trees are labeled NC for not canonical. All are rejected because of condition (2) of DEFINITION 6.77. If the tree structure of FIGURE 6.84 is continued through R_6, what is the first tree in preorder (relative to the tree of FIGURE 6.84) to be rejected because of condition (3) of DEFINITION 6.77?

In FIGURE 6.83 the trees were generated in breadth first order (DEFINITION 6.33). As indicated in FIGURE 6.84, the trees up to a fixed R_p can also be generated systematically in preorder (from depth first sequences, DEFINITION 6.31).

6.86 EXERCISE.

(1) How would you generate orbit diagrams of trees (not rooted)? Orbit diagrams of graphs? Orbit diagrams of connected graphs?
(2) Discuss the complexity of testing for canonicity in the sets of the form $B^{-1}(R_i)$ that arise in the orderly algorithm of FIGURE 6.83. More importantly, write a program to implement this orderly algorithm.

In FIGURE 6.65 all connected graphs with fixed vertex set are generated from all trees on that same vertex set with fixed root. These trees were generated there using the inverse Prüffer algorithm (INVPRU, procedure 6.21). These same trees can now be generated from the very different point of view of the orderly algorithm.

6.87 THE TREES OF FIGURE 6.65 OBTAINED FROM FIGURE 6.83. ALL ORBIT MEMBERS ARE ROOTED AT 0.

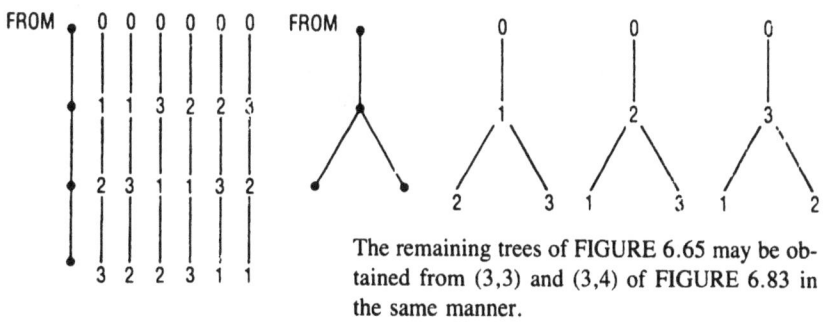

The remaining trees of FIGURE 6.65 may be obtained from (3,3) and (3,4) of FIGURE 6.83 in the same manner.

Figure 6.87

6.88 EXERCISE.

Complete the computation of the trees of FIGURE 6.87.

We have been applying the notion of a lineal spanning tree to the study of connected graphs. A graph that is connected in the sense that we have been discussing is sometimes called "1-connected." There are, as one might suspect from the terminology, notions of "2-connected," "3-connected," etc. Lineal spanning trees play an important role in the study of these higher orders of connectivity. Of particular importance are the notions of "2-connected" or "biconnected" graphs and "3-connected" or "triconnected" graphs. We shall study these cases in detail using lineal spanning trees. Before doing so, however, we consider briefly some other important ideas related to spanning trees.

6.89 DEFINITION.

Let G = (V,E) be a graph and let H = (V',E') be a subgraph. If V' = V then H is called a *spanning subgraph* of G.

We have already introduced the idea of a *spanning forest* in DEFINITION 6.47.

6.90 DEFINITION.

Let G = (V,E) be a graph. We define the *rank* of G, r(G), to be the number of edges in a spanning forest of G. Let e(G), v(G), and p(G) denote the number

47

of edges, vertices, and components of G, respectively. We define the *nullity* of G, $n(G)$, to be $e(G) - r(G)$.

The terms "rank" and "nullity" come from the linear algebraic approach to graph theory in which the numbers $r(G)$ and $n(G)$ turn out to be the rank and nullity of a certain naturally defined linear transformation. If G is a graph and H a subgraph, then we use the notation $H + e$ and $H - e$ to denote H with e added and deleted, respectively.

Let G be a connected graph, and let H be a spanning subgraph of G. We now define two basic algorithms, BREAK(H) and JOIN(H). In these algorithms we require that the edges of G be linearly ordered: $e_1, e_2, \ldots, e_n$.

6.91 *procedure* BREAK(H).

initialize $K := H$;
for $i := n$ *step* -1 *until* 1 *do*
   if e_i is in K and $r(K) = r(K - e_i)$ *then* $K := K - e_i$;
{At the end of the procedure K is BREAK(H).}

6.92 *procedure* JOIN(H).

initialize $K := H$;
for $i := 1$ *step* 1 *until* n *do*
   if e_i is in G but not in K and $n(K) = n(K + e_i)$ *then* $K := K + e_i$;
{At the end of the procedure K is JOIN(H).}

In EXERCISE 6.94(2), the reader is asked to show that, for e_i in K, $r(K) = r(K - e_i)$ if and only if e_i belongs to some cycle in K. Similarly, for e_i in G but not in K, $n(K) = n(K + e_i)$ if and only if the endpoints of e_i lie in different components of K. With these characterizations in mind we give an example of the execution of BREAK and JOIN in EXAMPLE 6.93.

6.93 EXAMPLE OF BREAK AND JOIN.

Graph G with edges 1,2,. . .,13;
subgraph H shown by solid lines

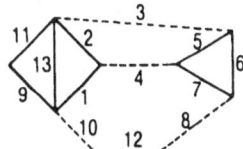

BREAK   JOIN

DELETE 13:   ADD 3:

DELETE 11:   ADD 8:

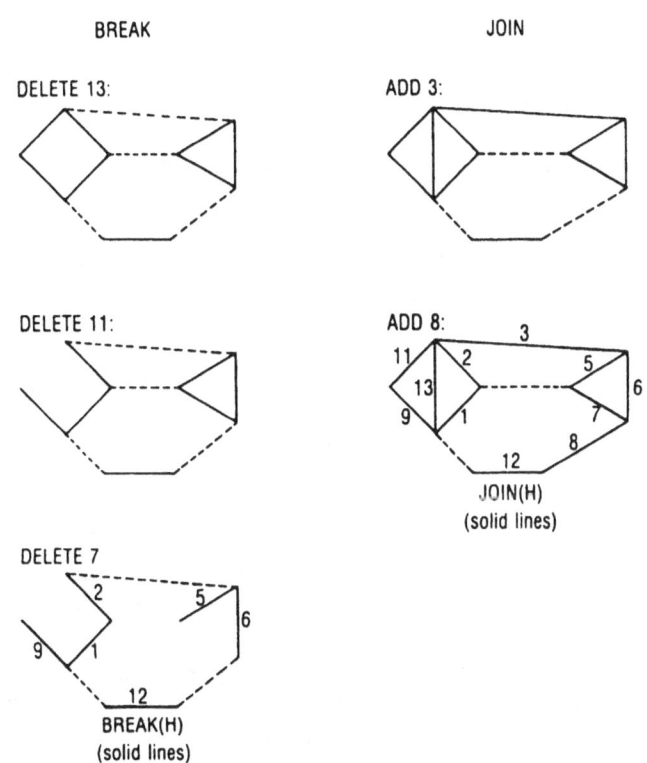

JOIN(H)
(solid lines)

DELETE 7

BREAK(H)
(solid lines)

Figure 6.93

6.94 EXERCISE.

(1) Let G = (V,E) be a graph. Prove that the number of edges in a spanning
    forest F of G is independent of the choice of F. This justifies the definition
    of r(G), the rank of G, given in DEFINITION 6.90. Show that in fact $r(G)$
    $= v(G) - p(G)$.

(2) Let K be a spanning subgraph of G. If e is an edge of K, show that $r(K) = r(K - e)$ if and only if e lies on some cycle of K. If e is in G but not in K, show that $n(K) = n(K + e)$ if and only if the endpoints of e lie in different components of K. (In other words, K + e has one less component than K.)

(3) Let G be a connected graph and let T be a spanning tree for G. If e is an edge of G but not an edge of T, show that there exists an edge f of T such that $(T - f) + e$ is a spanning tree of G. Characterize all such f. *Hint:* Consider the unique path in T joining the endpoints of e. Choose f from this path.

(4) Let $G = (V,E)$ be a connected graph with E linearly ordered: $e_1, e_2, \ldots, e_n$. Let $w: E \rightarrow R$ be a mapping from E to the real numbers R such that $i < j$ implies that $w(i) \leq w(j)$. For any subgraph K of G, define the *weight* of K to be $w(K) = \Sigma w(e)$ where the sum is over all edges of K. Let $D = (V, \phi)$ be the discrete spanning subgraph of G (D has no edges). Prove that JOIN(D) and BREAK(G) are spanning trees of minimum weight. The algorithms JOIN(D) and BREAK(G) are called "greedy algorithms" in this case because at each step they respectively add the minimal or discard the maximal relevant edge. A priori, it might be better to not always take the minimal weight edge (in order to improve the situation later on). But, this is not required in this case. What sort of extremal property characterizes JOIN(H) and BREAK(H) in the case of an arbitrary spanning graph H ("extremal" relative to the weight function w)?

As mentioned above, in addition to using lineal spanning trees to study the notion of connected graphs, we can use them to study the more refined notion of "2-connected" or "biconnected" graphs.

6.95 DEFINITION.

Let $G = (V,E)$ be a graph with $e, f \in E$. We say that e is "cycle equivalent" to f if $e = f$ or if e and f lie on the same cycle (DEFINITION 6.10). We write $e_{\hat{c}}f$ if e is cycle equivalent to f. This relation can be shown to be an equivalence relation (EXERCISE 6.96(1)). Let $E_1, \ldots, E_p$ denote the equivalence classes of the cycle equivalence relation. Let V_i denote the set of vertices belonging to edges of E_i. The subgraphs $G_i = (V_i, E_i)$, $i = 1, \ldots, p$, are called the *biconnected components* of G. G is *biconnected* if it has only one such component.

6.96 EXERCISE.

(1) Show that "cycle equivalence" is an equivalence relation (DEFINITION 1.2, CHAPTER 1). *Hint:* Transitivity needs some thought.

(2) Let $G_i = (V_i, E_i)$ and $G_j = (V_j, E_j)$ be two distinct bicomponents of G. Prove that $V_i \cap V_j$ has at most one element. If $x \in V_i \cap V_j$ then x is called an *articulation point* of G.

(3) Let G be a connected graph and let x, y, and z be three distinct vertices of G. Show that if every path from x to y must pass through z then z is an articulation point of G.

(4) Prove that a graph G is biconnected if and only if between every pair of distinct vertices there are two vertex disjoint paths (assume G has at least two edges).

The reader should note the general "global appearance" of the bicomponents of a connected graph G. FIGURE 6.97 shows a *possible* pattern of bicomponents in (a), an *impossible* (why?) pattern in (b). Note the "tree like" structure in FIGURE 6.97(a).

**6.97   GENERAL PATTERN FOR BICOMPONENTS.
LARGE DOTS ARE ARTICULATION POINTS.**

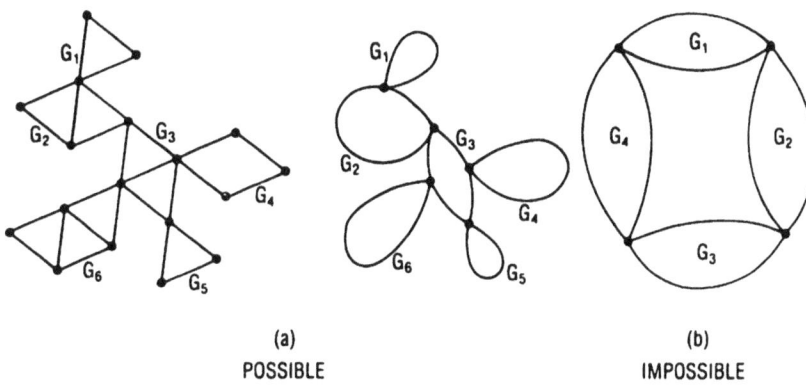

(a)
POSSIBLE

(b)
IMPOSSIBLE

Figure 6.97

In the second figure of (a) and in (b) the actual structure of the graph in the regions G_i is omitted.

6.98   THE BRIDGES OF A CYCLE IN A BICONNECTED GRAPH G.

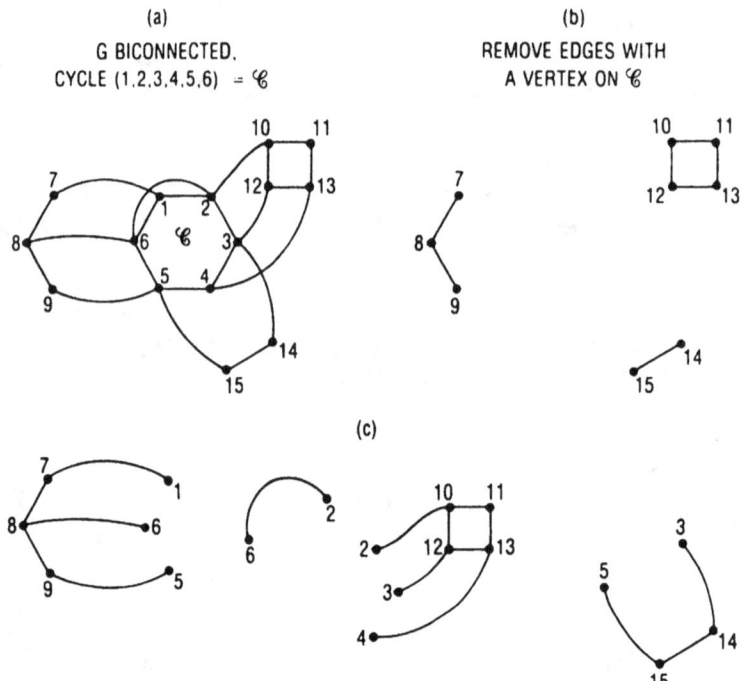

(a)
G BICONNECTED.
CYCLE (1,2,3,4,5,6) = $\mathscr{C}$

(b)
REMOVE EDGES WITH
A VERTEX ON $\mathscr{C}$

(c)

Restore simple bridges and restore edges incident on each component of (b) to obtain bridges of $\mathscr{C}$.

Figure 6.98

6.99   A MORE FUNDAMENTAL VIEW OF BRIDGES OF A CYCLE.

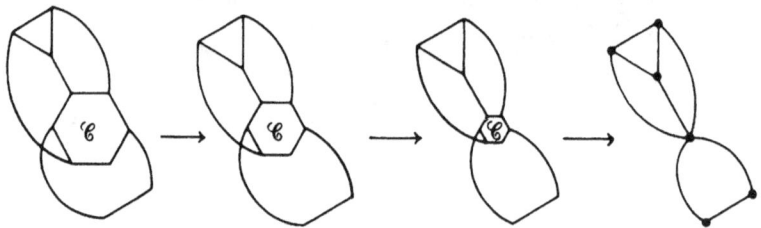

Start with a cycle $\mathscr{C}$ in a biconnected graph G. Let the cycle $\mathscr{C}$ shrink to a point. The edges of the biconnected components of the remaining graph are exactly the edges of the bridges in the original graph G.

Figure 6.99

We now develop the fundamental recursion for biconnected graphs. To do this, we need the notion of a *bridge* with respect to a cycle in a biconnected graph. This idea is illustrated in FIGURES 6.98 and 6.99 and EXERCISE 6.109.

6.100 DEFINITION.

Let H be a subgraph of a biconnected graph G. We define subgraphs of G called *bridges of G relative to* H. An edge of G that is not an edge of H is a bridge if both its vertices lie in H. Such a bridge is called a *simple bridge*. To define the other bridges, remove all vertices of H from G. Remove also all edges with a vertex in H (this includes all simple bridges and all edges of H). Let K be a component of the graph that is left. K together with all edges of G incident on K is a *bridge of G relative to* H. We are primarily interested in the case where H is a cycle $\mathscr{C}$ of G. (See FIGURE 6.98.)

The geometric idea behind the notion of a bridge relative to a cycle $\mathscr{C}$ is easy to visualize. Another way of presenting the same idea is shown in FIGURE 6.99. This approach to the notion of a bridge extends to structures called "matroids" and is thus a more fundamental way to look at bridges. We shall consider matroids in Chapter 10.

6.101 DEFINITION.

Let $\mathscr{C}$ be a cycle in a graph G and let B be a bridge relative to this cycle. Let $\mathscr{C}$ + B be the graph formed by taking the union of the edges in $\mathscr{C}$ and B. $\mathscr{C}$ + B is called a $\mathscr{C}$-bicomponent of the cycle.

The reader is asked to show in EXERCISE 6.109(2) that a $\mathscr{C}$-bicomponent is a biconnected graph if G is biconnected.

6.102 DEFINITION.

Let $\mathscr{C}$ + B be a $\mathscr{C}$-bicomponent. If e is an edge of $\mathscr{C}$ then $\mathscr{C}' = \mathscr{C} - e$ is called a *broken cycle* ($\mathscr{C}'$ is simply a path). The bicomponent of $\mathscr{C}'$ + B that contains B is called the *carrier* of B relative to the *broken cycle* $\mathscr{C}'$.

The reader is asked to show in EXERCISE 6.109(3) that B is in fact contained in a single bicomponent of $\mathscr{C}'$ + B. The carriers of the four bridges of the cycle $\mathscr{C}$ of FIGURE 6.98 are shown in FIGURE 6.103. In general, if $\mathscr{C}' = (v_1, v_2, . . ., v_k)$ where e = $\{v_1, v_k\}$, let v_i be the first vertex and v_j be the last vertex of $\mathscr{C}'$ that are vertices of B. Then the carrier of B in $\mathscr{C}'$ + B is simply B together with the edges and vertices of the path $(v_i, . . ., v_j)$.

6.103 THE CARRIERS OF THE BRIDGES OF FIGURE 6.98.
$\mathscr{C}' = (2,3,4,5,6,1)$

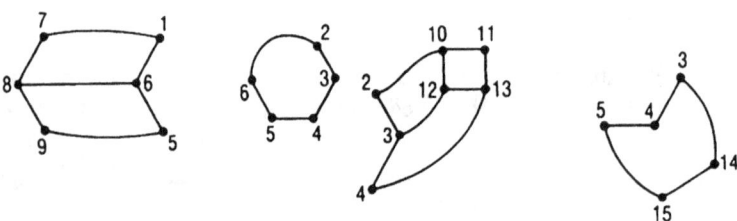

Figure 6.103

We are now prepared to describe the fundamental recursion for biconnected graphs. We do this by defining the *bicomponent tree* of a biconnected graph (DEFINITION 6.104).

6.104 DEFINITION.

A *bicomponent tree* of a biconnected graph is defined by the following rules:

(1) The root is the graph G together with a broken cycle $\mathscr{C}'$.
(2) Each internal vertex is a biconnected graph H together with a broken cycle $\mathscr{C}'$ in H.
(3) The sons of a vertex H are an ordered list of carriers (in H) of the bridges of H relative to $\mathscr{C}'$.
(4) A terminal vertex (leaf) of the bicomponent tree is an H that is itself a cycle. The broken cycle $\mathscr{C}'$ is omitted for a leaf.

A bicomponent tree of the graph of FIGURE 6.98 is shown in FIGURE 6.105. A bicomponent tree is an ordered rooted tree. There are, of course, many bicomponent trees associated with a graph G, depending on how the broken cycles are chosen and how the carriers are ordered. Two bicomponent trees for the complete graph K_5 are given in FIGURE 6.106. The complete graph K_n is a graph on n vertices in which every pair of vertices is an edge.

54

6.105   A BICOMPONENT TREE OF THE GRAPH OF FIGURE 6.98.

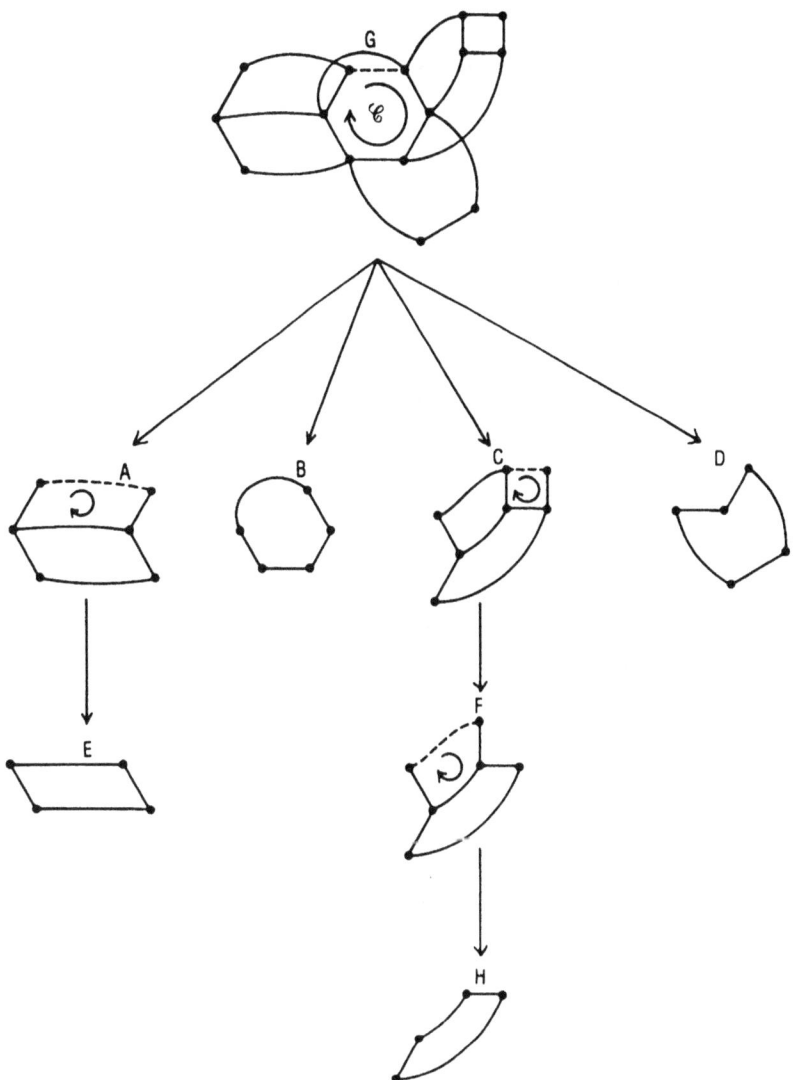

For vertex labels refer to FIGURES 6.98 and 6.103.

Figure 6.105

6.106   TWO BICOMPONENT TREES FOR K₅.

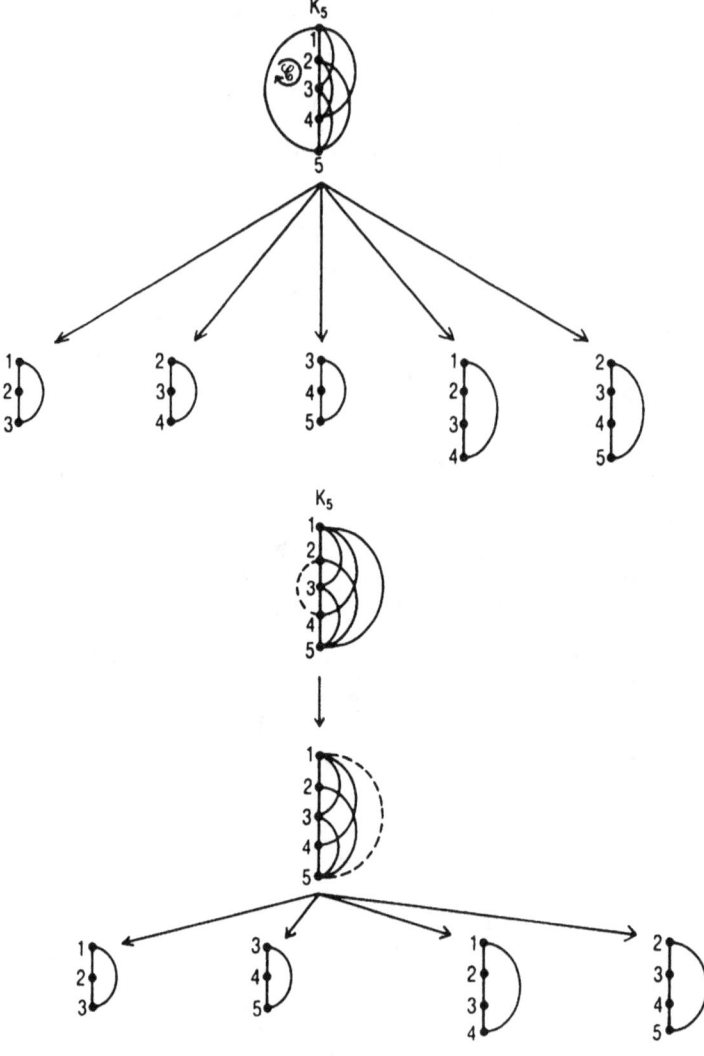

Figure 6.106

It is obvious from FIGURES 6.105 and 6.106 that there is much redundant information in the vertex labels of the bicomponent trees. Thus we introduce the idea of a *tree of cycles* or *cycle tree* of a biconnected graph.

6.107   DEFINITION.

Let $\mathcal{T}$ be the bicomponent tree of a biconnected graph G. Each internal vertex of $\mathcal{T}$ consists of a subgraph of G together with a broken cycle $\mathcal{C}'$ (obtained from

56

a cycle $\mathscr{C}$ by removing an edge). If each such vertex is replaced by the cycle $\mathscr{C}$ (of the broken cycle $\mathscr{C}'$), then the resulting tree is called a *cycle tree* or *tree of cycles* of the graph G.

FIGURE 6.108 shows the tree of cycles corresponding to the bicomponent tree of FIGURE 6.105.

6.108 THE TREE OF CYCLES CORRESPONDING TO FIGURE 6.105.

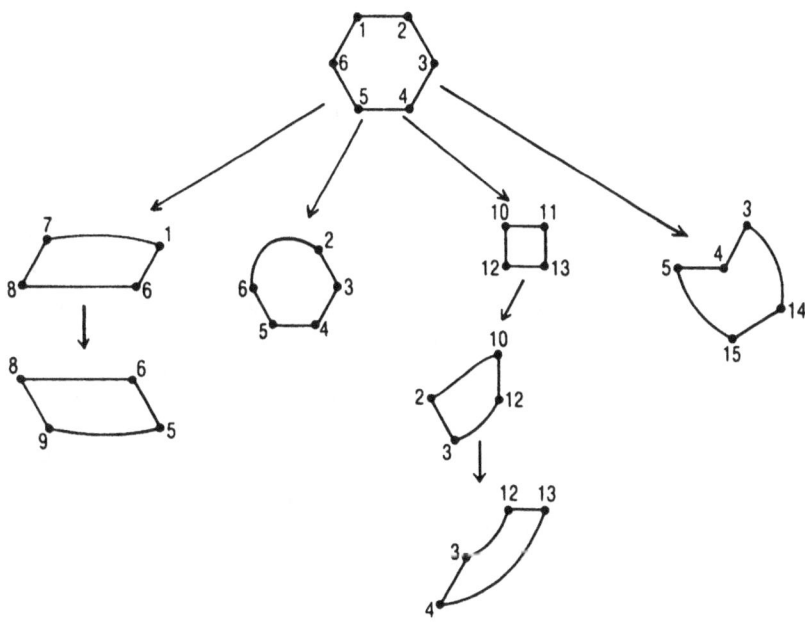

Figure 6.108

6.109 EXERCISE.

(1) Construct several nontrivial examples to illustrate the idea of *bridges, carriers, bicomponent tree,* and *cycle tree*.

(2) Let G be a biconnected graph and let $\mathscr{C}$ be a cycle in G. Prove that a "$\mathscr{C}$-bicomponent," $\mathscr{C} + B$ (see DEFINITION 6.101), is in fact a biconnected graph.

(3) Let $\mathscr{C} = (v_1, v_2, \ldots, v_k, v_1)$ be a cycle of a biconnected graph G. Let B be a bridge relative to $\mathscr{C}$. Let $\mathscr{C}' = (v_1, \ldots, v_k)$ and let v_i be the first and v_j be the last vertex of $\mathscr{C}'$ in B. Prove that $i = j$ never occurs (*Hint:* Use the fact that G is biconnected). Prove that the carrier of B relative to $\mathscr{C}'$ is B together with all vertices in the path $(v_i, \ldots, v_j)$. Call the path $(v_i, \ldots, v_j)$ the SPAN(B) relative to $\mathscr{C}'$.

(4) Let $\mathcal{T}$ be a bicomponent tree and $\mathcal{T}'$ be the corresponding cycle tree. Carefully describe data structures for representing $\mathcal{T}$ and $\mathcal{T}'$. Give an algorithm that constructs $\mathcal{T}$ from $\mathcal{T}'$ and discuss its complexity.

(5) Describe an algorithm that when given a biconnected graph G produces a bicomponent tree $\mathcal{T}$ and a cycle tree $\mathcal{T}'$. Discuss data structures and complexity.

(6) What happens if you try to construct the bicomponent tree for a graph G that is connected but not biconnected? Can you use this approach to give an algorithm for finding the biconnected components of a graph?

(7) Let H = (W,F) be a subgraph of G = (V,E). Define a relation $\sim$ (DEFINITION 1.2, CHAPTER 1) on $E - F$ by (i) $e \sim e$ for all $e \in E - F$ and (ii) $e \sim f$ if there is a path in G starting with e and ending with f and having no interior vertices in H. Prove that $\sim$ is an equivalence relation and the equivalence classes are the bridges of H (DEFINITION 6.100).

Before relating lineal spanning trees to the above ideas, there is one more general structure that we need to develop. Let G be a biconnected graph. Imagine G drawn in the plane as it is in the case of the graph of FIGURE 6.98(a). A drawing of a graph in the plane is called an *embedding* of the graph in the plane. If the graph can be embedded in the plane such that no two lines in the embedding touch (except possibly at a common vertex) then the graph is said to be *planar* and the embedding is called a *planar embedding*. The embedding of the graph of FIGURE 6.98(a) is clearly not planar although a planar embedding of this graph obviously exists. Let $\mathcal{C}$ be a cycle in a biconnected graph G. The cycle divides the plane into two regions, the bounded inner region and the unbounded outer region. Consider the list of bridges shown in FIGURE 6.98(c). Note that the first and second bridge cannot go on the same side of the cycle $\mathcal{C}$. Similarly, the third and fourth bridges must go on opposite sides of $\mathcal{C}$. We formalize this idea with the notion of a *bridge graph* of a cycle, DEFINITION 6.110.

6.110 DEFINITION.

Let G be a biconnected graph and let $\mathcal{C}$ be a cycle in G. Let BRGR($\mathcal{C}$) = ($\mathcal{U},\mathcal{E}$) denote a graph which we call the *bridge graph* of $\mathcal{C}$ in G. The vertex set $\mathcal{U}$ is the set of bridges of the cycle $\mathcal{C}$. A pair of bridges {X,Y} is in $\mathcal{E}$ if X and Y cannot be embedded on the same side of $\mathcal{C}$ in any planar embedding of X $\cup$ Y $\cup$ $\mathcal{C}$.

As an example, consider the cycle $\mathcal{C}$ of FIGURE 6.98(a) and its list of bridges shown in FIGURE 6.98(c). Call the bridges in this list, left to right, A, B, C, and D. The bridge graph BRGR($\mathcal{C}$) = ({A,B,C,D}, {{A,B}, {C,D}}).

Again, let $\mathcal{C}$ be a cycle in a biconnected graph G and assume that all $\mathcal{C}$-bicomponents are planar (i.e., all subgraphs of the form $\mathcal{C} \cup X$, X a bridge of $\mathcal{C}$, are planar). Then G itself is planar if it is possible to place the various bridges of $\mathscr{C}$ inside the outside of $\mathscr{C}$ in such a way that no lines cross in the embedding. Such a placement of bridges can be thought of as a labeling of the vertices of the bridge graph with the two labels I (for inside $\mathscr{C}$) and O (for outside $\mathscr{C}$). A moment's thought (draw a few pictures) suggests the following lemma.

6.111 LEMMA.

Let $\mathscr{C}$ be a cycle in a biconnected graph G and assume that each $\mathscr{C}$-bicomponent is planar. Then G is planar if the vertices of BRGR($\mathscr{C}$) can be labeled with the two symbols I and O such that the vertices of every edge of BRGR($\mathscr{C}$) have different labels.

The idea behind LEMMA 6.111 is that vertices of an edge of BRGR($\mathscr{C}$) must go on opposite sides of $\mathscr{C}$ in any embedding and thus must have opposite labels. LEMMA 6.111 requires the intuitively obvious result that a closed simple curve in the plane has an inside region and an outside region. This result is known as the "Jordan Curve Theorem" in topology. In general, a graph G is called "2-colorable" or "bipartite" if its vertices can be labeled with two symbols (such as I and O) such that the two vertices of every edge are labeled differently. In EXERCISE 6.118(1) the reader is asked to show that a graph is bipartite if and only if it contains no cycle of odd length (odd number of edges).

The ideas in LEMMA 6.111 and the previous paragraph provide the basis for a number of algorithms to test whether or not a graph is planar. The obvious idea is to start with a biconnected graph G and a cycle $\mathscr{C}$ in G. If the graph being tested is not biconnected to begin with, then find its biconnected components and work with them, as a graph is planar if and only if each of its biconnected components is planar. Recursively, determine whether or not each of the $\mathscr{C}$-bicomponents is planar. If any $\mathscr{C}$-bicomponent is not planar then the graph G is not planar (why?). If all of the $\mathscr{C}$-bicomponents are planar then the graph G is planar if and only if BRGR($\mathscr{C}$) is bipartite. There is a slight hitch here, however! If $\mathscr{C}$ has only one bridge then the $\mathscr{C}$-bicomponent is G itself. One might instead work with the carrier of the one bridge. The carrier might be planar but G nonplanar, however. These complications are illustrated in FIGURE 6.112. These problems turn out to be minor ones. The reader is asked to consider them in EXERCISE 6.118(3).

6.112   A NONPLANAR GRAPH WITH ONE 𝒞-BICOMPONENT AND PLANAR CARRIER.

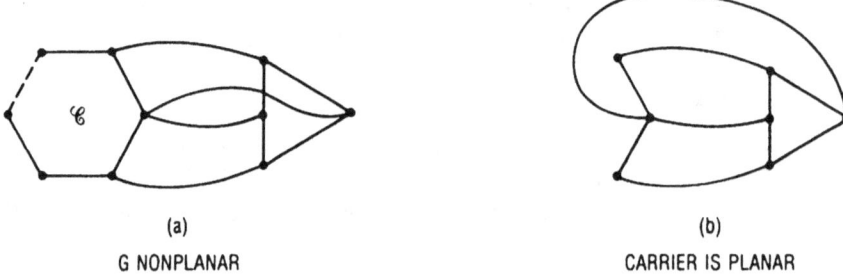

(a)

G NONPLANAR

(b)

CARRIER IS PLANAR

Figure 6.112

We shall now give a prototypic planarity algorithm based on the bicomponent tree of a biconnected graph G. This algorithm is intended as a rough guide to planarity testing (and is hence called the "sloppy planarity" algorithm). A more careful discussion will be given later on. Let $\mathcal{T}$ be a bicomponent tree of a biconnected graph G (see FIGURE 6.105, for example). Recall that each vertex X of $\mathcal{T}$ is a subgraph of G together with a broken cycle $\mathcal{C}'_X$. Recalling the notation for vectors, 6.42, let $(X,Y) \leftarrow L$, where L is a list of edges of $\mathcal{T}$, denote the operations of deleting the first edge of L and calling it (X,Y). The edge (X,Y) is directed away from the root (X is closer to the root of $\mathcal{T}$ than Y). Initially, we let L be the postorder list of edges of $\mathcal{T}$ (DEFINITION 6.31), designated by POSE($\mathcal{T}$). For each vertex X of $\mathcal{T}$ we shall construct an embedding, EMBED(X). Initially, EMBED(X) is just the cycle $\mathcal{C}_X$.

6.113   *procedure* SLOPPY PLANARITY TEST.

begin
initialize L: = POSE($\mathcal{T}$) ($\mathcal{T}$ a bicomponent tree of G)
          BRGR($\mathcal{C}_X$): = (ϕ,ϕ) for all vertices X of $\mathcal{T}$; (see DEFINITION 6.110)
          EMBED(X): = $\mathcal{C}_X$ embedded in the plane;
while L $\neq \phi$ *do*
  begin
    $(X,Y) \leftarrow L$;
    if $\mathcal{C}_X \cup$ Y is planar *then* add Y to BRGR($\mathcal{C}_X$) *else* STOP, G IS NON-
      PLANAR;
    if BRGR($\mathcal{C}_X$) is bipartite *then* add Y to EMBED(X) *else* STOP, G IS
      NONPLANAR;
  end
G IS PLANAR AND EMBED(G) IS AN EMBEDDING OF G
end

The SLOPPY PLANARITY TEST, *procedure* 6.113, is intended as an intuitive guide for "pencil and paper" execution only. The exact manner that Y is added to the bridge graph, BRGR($\mathscr{C}_x$), is not specified, the test to see if BRGR($\mathscr{C}_X$) is bipartite is not specified, and how Y is added to the embedding, EMBED(X), is not specified. We now give an example of the "execution" of the SLOPPY PLANARITY TEST in the example of FIGURE 6.105. We refer to the labels G,A,B,. . . of the vertices of the bicomponent tree of FIGURE 6.105.

6.114 SLOPPY PLANARITY TEST APPLIED TO FIGURE 6.105.

L: = (A,E),(G,A),(G,B),(F,H),(C,F),(G,C),(G,D).
(A,E) ← L: $\mathscr{C}_A \cup$ E is planar, so add E to BRGR($\mathscr{C}_A$) to get BRGR($\mathscr{C}_A$): = ({E},ϕ).
  BRGR($\mathscr{C}_A$) is bipartite so add E to EMBED(A) to get

(G,A) ← L: $\mathscr{C}_G \cup$ A is planar, so add A to BRGR($\mathscr{C}_G$) to get BRGR($\mathscr{C}_G$): = ({A},ϕ).
  BRGR($\mathscr{C}_G$) is bipartite so add A to EMBED(G) to get

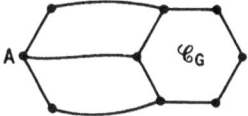

(G,B) ← L: $\mathscr{C}_G \cup$ B is planar, so add B to BRGR($\mathscr{C}_G$) to get ({A,B},{{A,B}}).
  BRGR($\mathscr{C}_G$) is bipartite so add B to EMBED(G) to get

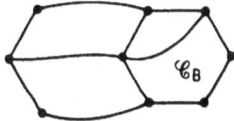

(F,H) ← L: $\mathscr{C}_F \cup$ H is planar, so add H to BRGR($\mathscr{C}_F$) to get ({H},ϕ).
  BRGR($\mathscr{C}_F$) is bipartite so add H to EMBED(F) to get

61

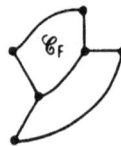

(C,F) ← L: $\mathscr{C}_C \cup$ F is planar, so add F to BRGR($\mathscr{C}_C$) to get ({F},φ).
    BRGR($\mathscr{C}_C$) is bipartite so add F to EMBED(C) to get

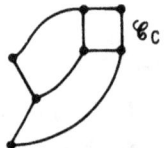

(G,C) ← L: $\mathscr{C}_G \cup$ C is planar, so add C to BRGR($\mathscr{C}_G$) to get ({A,B,C},{{A,B}}).
    BRGR(G) is bipartite so add C to EMBED(G) to get

(G,D) ← L: $\mathscr{C}_G \cup$ D is planar, so BRGR($\mathscr{C}_G$):= ({A,B,C,D}, {{A,B}, {C,D}}).
    BRGR($\mathscr{C}_G$) is bipartite so add D to EMBED(G) to get

EMBED(G)

L = φ so STOP, G IS PLANAR

The reader is asked to explore some additional properties of the SLOPPY
PLANARITY TEST in EXERCISE 6.118. There is one very simple test of
planarity that works for some graphs. This test is stated in THEOREM 6.115
and proved in EXERCISE 6.118(4).

6.115 THEOREM.

If G = (V,E) is a connected planar graph with at least three edges then $|E| \leq$
$3|V| - 6$. If in addition G is bipartite then $|E| \leq 2|V| - 4$.

As an example, consider K_5, the complete graph on five vertices, and $K_{3,3}$, the complete bipartite graph on six vertices partitioned into two groups of 3. These graphs are shown in FIGURE 6.116.

6.116 K_5 and $K_{3,3}$.

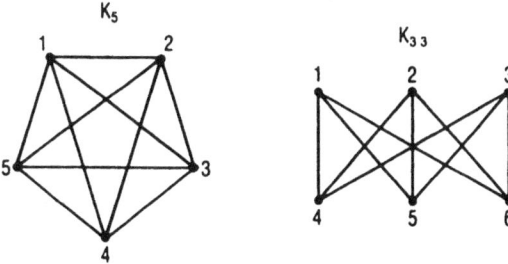

Figure 6.116

For K_5 the inequality $|E| \le 3|V| - 6$ becomes $10 \le 9$ and for $K_{3,3}$ the inequality $|E| \le 2|V| - 4$ becomes $9 \le 8$. Thus neither graph is planar.

Theorem 6.115 follows from a famous result of Euler that for any *connected* planar embedding of a graph $|V| - |E| + |R| = 2$ where $|R|$ is the number of "regions" in the embedding. It is very easy to see why this result is true. Consider the embedding of FIGURE 6.117. The regions are labeled r_1, r_2, r_3, and r_∞. The latter region is called the "unbounded region" (it would be just another bounded region if the embedding were on a sphere instead of the plane). The edge p of this embedding is called "pendant" because it has a vertex of degree 1. The edge e is not pendant and neither is the edge f. The edge f is called an "isthmus" as its removal disconnects the graph. If an edge such as e is removed (i.e., an edge that is not pendant and not an isthmus), then an embedding such as that shown in FIGURE 6.117(b) results. Call this a TYPE I deletion. Notice that a TYPE I deletion reduces $|E|$ by 1 and $|R|$ by 1. Thus a TYPE I deletion does not change the sum $|V| - |E| + |R|$. Also, the graph remains connected. A TYPE II deletion consists of removing a pendant edge and its vertex of degree 1 (see FIGURE 6.117(c)). A TYPE II deletion decreases $|V|$ by 1 and $|E|$ by 1 so again the sum is not changed. The reader can easily see by trying some examples that a series of TYPE I and TYPE II deletions will always reduce a connected planar embedding to the graph with one vertex and no edges. This graph (when embedded in the plane as a dot!) has one region, the unbounded region, and hence $|V| - |E| + |R| = 2$ for this embedding and thus for the original embedding also.

6.117 TYPE I AND TYPE II DELETIONS.

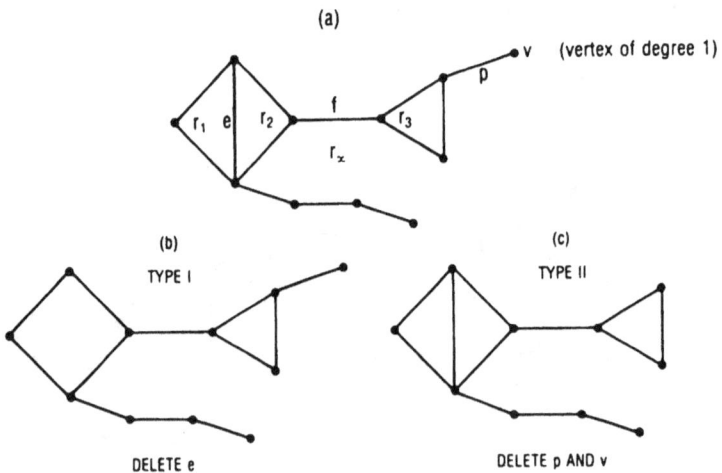

Figure 6.117

6.118 EXERCISE.

(1) A graph $G = (V,E)$ is *bipartite* if V can be partitioned into two sets $\{V_1, V_2\}$ such that $e \in E$ implies that one vertex of e is in V_1 and the other is in V_2. G is *2-colorable* if the elements of V can be labeled with two symbols such that the two vertices of any edge have different symbols. Prove that G is bipartite if and only if G is 2-colorable. Prove that G is bipartite if and only if G has no cycles of odd length.

(2) Use the SLOPPY PLANARITY TEST, *procedure* 6.113, to show that K_5 and $K_{3,3}$ (FIGURE 6.116) ARE NOT PLANAR.

(3) The SLOPPY PLANARITY TEST is a recursive procedure which, for each edge (X,Y) in the bicomponent tree, looks at the $\mathscr{C}_X$-bicomponent $\mathscr{C}_X \cup Y$. This is not a true recursion (reduction to simpler cases) if $\mathscr{C}_X$ has Y as its only bridge (see FIGURE 6.112). In this case, $X = \mathscr{C}_X \cup Y$. How can this problem be avoided to make the SLOPPY PLANARITY TEST a little less sloppy? (*Hint:* When can the cycle be changed to one that has at least two bridges?)

(4) Use the Euler relation, $|V| - |E| + |R| = 2$, to prove THEOREM 6.115. (*Hint:* Imagine a list ER of all pairs (e,r) where e is an edge "on the boundary" of region r. This means that all points of the edge are in contact with the region or "the closure of the region" in topological terms. Any edge can be paired with at most two regions. A pendant edge is paired with only one region. Under our assumptions a fixed region r must be paired with at least three edges or at least four if the graph is bipartite. Thus, in the

64

general case, we must have $3|R| \leq |ER| \leq 2|E|$ or $4|R| \leq |ER| \leq 2|E|$ in the bipartite case.)

It is a remarkable fact that by using lineal spanning trees and carefully chosen data structures the SLOPPY PLANARITY TEST can be converted into a planarity test that is (using the direct access model) linear in the number of vertices of the graph. This means that there is an algorithm and a constant c such that given any graph $G = (V,E)$, the algorithm decides in time less than or equal to $c|V|$ whether or not the graph is planar. The planar embedding, if it exists, can also be specified. Moreover, the algorithm is not a purely theoretical result of asymptotics and complexity arguments. The constant c is reasonable and the resulting algorithm appealing in practical terms. We shall, in CHAPTER 7, TOPIC I, develop in detail, using numerous examples, the basic ideas of this algorithm. One can make statements about the importance of planarity testing in applied problems (design of printed circuits, etc.) but this is risky ground. As the saying goes, "There is more than one way to skin a cat!." A better reason for us to look carefully at the linear time planarity algorithm is that it was a difficult problem to solve in the first place. Moreover, it deals with a problem of intrinsic mathematical interest that had been considered from different points of view by a number of good mathematicians. Thus it represents an excellent case study of a hard problem solved from the algorithmic point of view. The reader content with the SLOPPY PLANARITY TEST can skip to the discussions of triconnectedness or matroids, TOPICS III or IV, Chapter 9 or 10.

Unit 7

Depth First Search and Planarity

We begin with a quick review of certain basic ideas presented in Chapter 6. As in DEFINITION 6.6, let I_z denote the set of all edges of $G = (V,E)$ incident on $z \in V$. A graph G will be called *ordered* (DEFINITION 6.6) if for each $z \in V$ a linear order is specified on I_z. Technically, an ordered graph G is a triple $(V,E,\{(I_z, \leqslant_z): z \in V\})$ where the last component is a family of linear orders. We avoid this cumbersome notation. As in DEFINITION 6.28, let A_z denote the vertices adjacent to z. An ordering on A_z produces an ordering on I_z. A connected acyclic graph $T = (V,E)$ is a *tree* (DEFINITION 6.12). If in addition to T we specify a particular vertex $x \in V$ then the triple (V,E,x) is a *rooted tree* with *root* x (DEFINITION 6.23). If we delete x from V and delete all incident edges I_x from E, then the remaining graph (if not empty) will decompose into one or more connected components, themselves trees, which we call the *principal subtrees of the root* x (DEFINITION 6.28). If $T' = (V',E')$ is such a subtree then there is a unique $x' \in V'$ such that $\{x,x'\} \in I_x$. We make the convention that this x' is the *root of* T'. Any rooted tree $T = (V,E,x)$ has associated with it a natural *directed* rooted tree: to each edge $\{x,x'\}$ in I_x we assign the directed edge (x,x'), (i.e., directed "away from" the root). Recursively, we direct the edges of the subtrees of the root x to obtain a directed rooted tree which we also denote by $T = (V,E,x)$ (see DEFINITION 6.25). Using this convention we may regard any rooted tree as a directed graph, and we do so whenever it is convenient. When we speak of a *directed rooted tree* we mean directed in the above sense unless otherwise specified. The most basic structure for describing algorithms is the *ordered (directed) rooted tree*, ORTR (FIGURE 6.27, DEFINITION 6.28, and paragraph following DEFINITION 6.28). An ORTR is shown in FIGURE 7.1. The ordering on A_z is first the father of z, and then the sons, left to right (see NOTATION FOR TREES, 6.41).

Given an ordered (directed) rooted tree $T = (V,E,x)$ let $T_1 = (V_1,E_1,x_1),\ldots,T_k = (V_k,E_k,x_k)$ be the rooted subtrees of x in order. We define a sequence of edges DFE(T) ("depth first" sequence of edges) and a sequence of vertices DFV(T) ("depth first" sequence of vertices) as follows (see DEFINITION 6.30 also):

(1) If $V = \{x\}$ then DFV(T) = x and if $E = \{(x,y)\}$ then DFE(T) = (x,y),(x,y).

(2) *Otherwise*:

DFE(T) = $(x,x_1),DFE(T_1),(x,x_1),\ldots,(x,x_k),DFE(T_k),(x,x_k)$

DFV(T) = $x,DFV(T_1),x \ldots x,DFV(T_k),x.$

These lists, for the directed ordered rooted tree of FIGURE 7.1(a) are given in FIGURE 7.1. The intuitive interpretation of these lists comes from FIGURE 7.1(b) as follows: Follow the arrows around the tree as shown. To construct DFE(T) list each edge as it is traversed (forward or backward). To construct DFV(T) start with the root, *a*, and list each vertex when it is encountered (thus "*a*" is listed four times in FIGURE 7.1).

The reader should recall the definitions of a *spanning forest* and *spanning tree* given in DEFINITION 6.47.

7.1 MORE DEPTH-FIRST SEQUENCES.

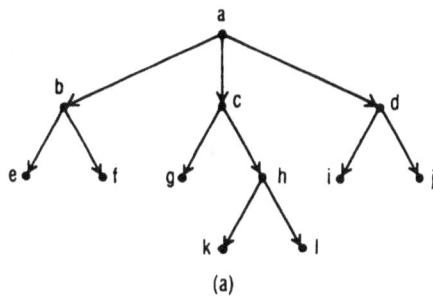

(a)

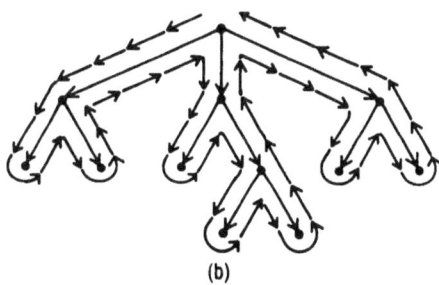

(b)

Figure 7.1

DFE(T) = (a,b),(b,e),(b,e),(b,f),(b,f),(a,b),(a,c),(c,g),(c,g),(c,h),(h,k)
          (h,k),(h,l),(h,l),(c,h),(a,c),(a,d),(d,i),(d,i),(d,j),(d,j),(a,d)

DFV(T) = a,b,e,b,f,b,a,c,g,c,h,k,h,l,h,c,a,d,i,d,j,d,a

Given any rooted tree $T = (V',E',x')$, we say that two vertices $w_1,w_2 \in V'$ are lineal with respect to T if $w_1 = w_2$ or they lie on the same directed path of T. If the path leads from w_1 to w_2 (or $w_1 = w_2$) we say that w_1 is an *ancestor* of w_2, and w_2 is a *descendant* of w_1. A rooted spanning tree $T = (V,E_T,x)$ is a *lineal* spanning tree for $G = (V,E)$ if $\{s,t\} \in E - E_T$ implies that s and t are lineal in T (DEFINITION 6.49, EXAMPLE 6.50). Given any connected graph $G = (V,E)$ and any $x \in V$, then there always exists a lineal spanning tree $T = (V,E_T,x)$ for G. This result was proved in THEOREM 6.52 and an algorithm given in *procedure* 6.56. An alternative proof was suggested in EXERCISE 6.54 and EXAMPLE 6.55.

It is worthwhile to understand both points of view, so we now discuss briefly the latter. Let $\{x,y\} \in I_x$. Referring to FIGURE 7.2, let $\bar{V}_y$ denote the set of all $s \in V$ for which no simple path from s to x includes y. Let $V_y = V - \bar{V}_y$. We assume $x \in \bar{V}_y$. Let G_y denote the restriction of G to V_y (all $\{s,t\}$ of G with $\{s,t\} \subseteq V_y$) and $\bar{G}_y$ the restriction of G to $\bar{V}_y$. From the definition of $\bar{V}_y$, the only possible edges of G not in either G_y or $\bar{G}_y$ are edges of the form $\{x,z\}$ for $z \in V_y$. Recursively we may construct lineal spanning trees T_y for G_y and $\bar{T}_y$ for $\bar{G}_y$. Adding the edge $\{x,y\}$ to T_y union $\bar{T}_y$ clearly produces a lineal spanning tree T for G rooted at x. Let $T = (V,E_T,x)$ and let $E_B = E - E_T$. The subgraph of G with edge set E_B will be called B. The edges E_B are called "backedges" or "fronds" and the edges E_T are called "tree edges." In general, the edges of $E - E_T$, T a spanning tree are called the *chords* of T in G (see DEFINITION 6.47). The term "backedges" is reserved for the chords of a *lineal* spanning tree. Henceforth we shall regard T and B as directed graphs with the following conventions:

(1) Edges of T are directed away from the root in the canonical fashion.
(2) If $(s,t) \in E_B$ then s is a descendant of t in T.

An example is given in FIGURE 7.2(b).

7.2 BASIC RECURSION FOR LINEAL SPANNING TREES.

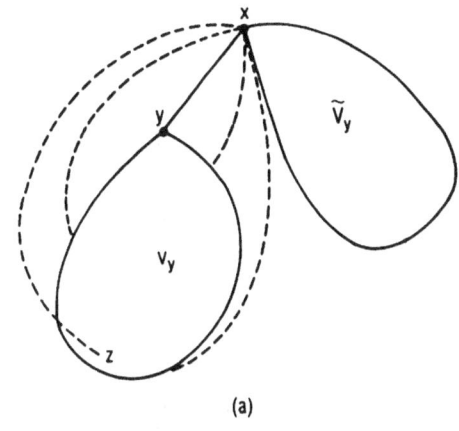

(a)

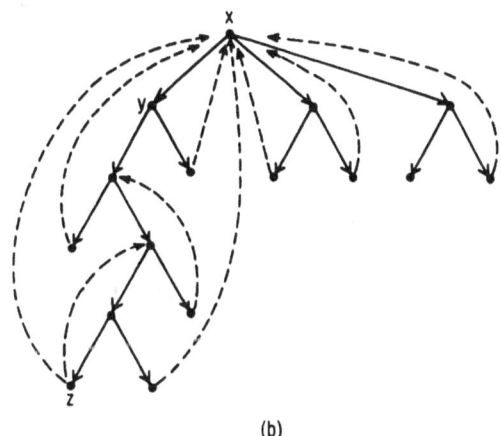

(b)

Solid arrows are edges of E_T, dashed arrows are edges of E_B.

Figure 7.2

This recursive description of the lineal spanning tree T and directed graph of backedges B (FIGURE 7.2) suggests a simple recursive algorithm for constructing T and B. First some preliminary remarks. Given an ordered directed rooted tree $T = (V, E_T, x)$ we define two functions FATHER and PREORDER on V. If $t \neq x$ then FATHER(t) is the unique $s \in V$ such that $(s, t) \in E_T$. The *preorder* linear order on V is the order specified by the sequence of first occurrences of the elements of V in the list DFV(T). For example, in the sequence DFV(T) of FIGURE 7.1 we extract just those entries z that represent a first occurrence of z to get a b e f c g h k l d i j, which is the preorder linear order of V for this example. The function that to each $z \in V$ assigns its position in preorder on V will be called PREORDER. Thus, for example in FIGURE 7.1

$$PREORDER = \begin{pmatrix} a\ b\ c\ d\ e\ f\ g\ h\ i\ j\ k\ l \\ 1\ 2\ 5\ 10\ 3\ 4\ 6\ 7\ 11\ 12\ 8\ 9 \end{pmatrix}.$$

Similarly, we define POSTORDER where "first occurrences" is replaced by "last occurrences." We refer the reader also to DEFINITION 6.30 and the related discussion.

Procedure 7.3 starts with a given ordered connected graph $G = (V,E)$ and $x \in V$. We wish to produce a lineal spanning tree $T = (V,E_T,x)$, DFE(T), the graph $B = (V_B,E_B)$, and the functions FATHER, PREORDER, and POSTORDER. Both T and B are produced as directed graphs. $PT = (PV,PE_T,x)$ and $PB = (PV_B,PE_B)$ are used to store the parts of T and B as they are being constructed. "CARD" denotes cardinality (number of elements) and v is the last vertex added to PV (maximum in PREORDER on PV). For each $v \in V$, A(v) initially is the list of vertices of G adjacent to v. A(v) is ordered by the order on edges I_v incident on v in G. DFE(T) is a list of edges (the depth-first sequence of edges of T when *procedure* 7.3 terminates). POST is an integer.

7.3 *procedure* **CONSTRUCT T AND B DESCENDANT TO v.**

begin
    while A(v) $\neq$ ϕ *do*
      begin
            w $\leftarrow$ first element of A(v) and delete w from A(v);
        if w $\in$ PV and PREORDER(w) $<$ PREORDER(FATHER(v)) *then*
            add (v,w) to PE_B and v and w to PV_B;
        if w $\notin$ PV *then*
        begin
        add (v,w) to PE_T and to DFE(T);
        add w to PV and set CARD(PV) $\leftarrow$ CARD(PV) + 1;
        FATHER(w) $\leftarrow$ v;
        PREORDER(w) $\leftarrow$ CARD(PV);
        CONSTRUCT T AND B DESCENDANT TO w;
        end
      end
    POST $\leftarrow$ POST + 1 and POSTORDER(v) $\leftarrow$ POST;
    if v $\neq$ x *then* add (FATHER(v),v) to DFE(T);
end

To construct T and B we set $PT = (\{x\},\phi,x)$, $PB = (\phi,\phi)$, PREORDER(x) = 1, POST $\leftarrow$ 0, CARD = 1, DFE(T) = ϕ and execute CONSTRUCT T AND B DESCENDANT TO x. At the end T = PT, B = PB.

7.4 COMMENTS.

(1) One can show that the above algorithm can be implemented in linear time in $|E|$ for connected graphs, $|E| + |V|$ in general. See EXERCISE 7.7(2).

(2) The statement "*if* w $\in$ PV and PREORDER(w) < PREORDER (FATHER(v)) *then*. . ." tests that (v,w) is a backedge not yet added to PB. The condition, if true, assures us that w is an ancestor of v other than the father of v.

(3) If G = (V,E) is planar, then it is a consequence of Euler's theorem on planar graphs that $|E| \leq 3|V| - 6$ (THEOREM 6.115, EXERCISE 6.118(4)). Suppose G is nonplanar and we apply the above procedure to each component G' of G, keeping a count of all edges in PB' $\cup$ PT' at each stage of the computation for that component. If this count ever gets to 3 CARD(PV') $- 5$ then we have a connected nonplanar subgraph of that component. If not, we keep the component just inspected and go to the next component. In this way we can determine a nonplanar subgraph $\tilde{G} = (\tilde{V},\tilde{E})$ of G with $|\tilde{E}| \leq 3 |V| - 5$. This observation is sometimes useful in dealing with nonplanar graphs, for it reduces the general case to that where the number of edges is bounded by a fixed linear function of the number of vertices. This reduction can be carried out in linear time in $|\tilde{V}|$ provided one assumes that the graph G is already given and represented by a suitable data structure (such as that of FIGURE 6.57). Of course, if the graph G has to be rewritten or scanned completely prior to the computation then all edges of E must be dealt with and linearity is destroyed in the general case. Even in this case, however, the reduction might be useful in reducing the constants involved in the computation.

In DEFINITION 6.95, we defined the notion of "cycle equivalence" of edges of a graph. That is to say, we defined a relation $\sim$ on E by $e_1 \sim e_2$ if $e_1 = e_2$ or if there is an elementary (not self-intersecting) cycle of G containing both e_1 and e_2. Clearly $\sim$ is reflexive and symmetric. Less clear, but easily shown, is transitivity: $e_1 \sim e_2$ and $e_2 \sim e_3$ implies $e_1 \sim e_3$ (EXERCISE 6.96(1)). Thus, $\sim$ is an equivalence relation. The subgraphs of G determined by the equivalence classes are called the *biconnected components* ("bicomponents") of G, (DEFINITION 6.95). If $G_1 = (V_1,E_1)$ and $G_2 = (V_2,E_2)$ are two biconnected components of G then either $V_1 \cap V_2 = \phi$ or $|V_1 \cap V_2| = 1$. If $x \in V_1 \cap V_2$ then x is called an *articulation point* of G (EXERCISE 6.96(2,3)).

We indicate how one finds the biconnected components of a graph efficiently. Given an ordered connected graph G = (V,E) and $x \in V$, construct the ordered directed graphs $T = (V,E_T,x)$ and $B = (V_B,E_B)$ as above.

7.5 DEFINITION.

For any e = (a,b) in E_T define LOW1(e) (the "first lowpoint" of e) to be the minimum of

(1) PREORDER(b)

and

(2) The minimum value of PREORDER(t) over all $(s,t) \in E_B$ with s a descendant of b (including $s = b$) in T.

If we exclude from (2) above all vertices t such that PREORDER(t) = LOW1(e) and again minimize over (1) and (2) we obtain LOW2(e) (the "second lowpoint" of e). Note that PREORDER(b) $\geq$ LOW2(e) > LOW1(e) unless LOW1(e) = PREORDER(b).

The LOW1 function may be computed recursively by examining E_T in postorder. Given $f = (p,q)$ in E_T, the edges $f_1,\ldots,f_s$ incident on q (directed away from q) are less than f in postorder. We assume inductively that LOW1 has been computed for all $f' < f = (p,q)$ in postorder. Let $m_1 = \text{MIN}\{\text{LOW1}(f_i); i = 1,\ldots,s\}$ and $m_2 = \text{MIN}\{\text{PREORDER}(z): (q,z) \in E_B\}$ (set $m_1 = \infty$ or $m_2 = \infty$ if the corresponding set is empty). Then LOW1(f) = $\text{MIN}\{m_1,m_2,\text{PREORDER}(q)\}$. It is easily seen that LOW1 can be computed in linear time in $|E|$. Similarly LOW2 can be computed in linear time in $|E|$. If b is not assumed connected, these computations are linear in $|V| + |E|$. *For e = $(a,b) \in E_B$ we adopt the convention that* LOW1(e) = PREORDER(b) *and* LOW2(e) = PREORDER(a).

Let $e = (a,b)$ be an edge of E_T. We say that an edge $e' = (a',b')$ is *descendant to e in* T if $e' = e$ or a' is a descendant of b (as defined above for vertices).

7.6 COMPUTING BICOMPONENTS.

To compute the biconnected components ("bicomponents") of a connected graph $G = (V,E)$, scan E_T in postorder until the first edge $f = (p,q)$ is found with LOW1(f) $\geq$ PREORDER(p) (if p is the root of T this inequality must hold). The set of tree edges of the first bicomponent is the set of all f' in E_T descendant to f (a consecutive sequence of edges prior to f in postorder). The set of backedges of the first bicomponent consists of all edges of E_B incident on vertices p' descendant to q. Delete this list of edges from $E_T \cup E_B$ and continue through E_T in postorder until the next edge $f = (p,q)$ is found such that LOW1(f) $\geq$ PREORDER(p), etc.

7.7 EXERCISE.

(1) In *procedure* 6.56 we gave an algorithm for computing a lineal spanning tree of a graph G based on the ideas of the proof of THEOREM 6.52. The algorithm described in *procedure* 7.3 is based on the recursive description (EXERCISE 6.54, FIGURE 6.55, FIGURE 7.2). Modify *procedure* 6.56 to compute the same structures as *procedure* 7.3.

(2) Discuss the complexity of *procedure* 7.3 or your alternative version based on *procedure* 6.56. Describe data structures.

(3) Write a more formal description of the algorithm for computing the bicomponents of a graph outlined in 7.6, above (pidgin ALGOL or, better yet, a
working program). Prove that *procedure* 7.6 actually works. (*Hint*: Suppose
that e = (a,b) ≠ f is an edge descendant to f = (p,q), where f is as described
in *procedure* 7.6. By construction, of f, LOW1(e) < PREORDER(a). Thus,
if a′ is the FATHER(a) then (a′,a) lies on a cycle with (a,b). Referring to
the edge (a′,a) as the *father* of the edge (a,b), we have shown that each such
edge e descendant to f is cycle equivalent to its father and thus, by transitivity,
equivalent to f. The condition LOW1(f) ≥ PREORDER(p) clearly implies
that p is an articulation point of G and thus we see that the set of edges
removed is indeed a bicomponent of G. The fact that the set of tree edges
removed defines a consecutive sequence in postorder, POSE(T), gives the
basis for an inductive proof.)

Thus far we have, given a connected, ordered graph G = (V,E), discussed the
decomposition of G into an ordered, rooted, directed spanning tree T = (V,E_T,x)
and a directed graph or backedges B = (V_B,E_B). The order on T and B is that
inherited from G. *We assume henceforth that G is biconnected.* We have discussed the functions PREORDER, LOW1, and LOW2. We now consider another
ordering of T and B that greatly facilitates subsequent computations. Using the
notation X(*statement*) = 0 if *statement* = false and 1 if *statement* = true, we
associate with each f = (p,q) ∈ E_T ∪ E_B the triple (PREORDER(p), LOW1(F),
X(LOW2(f) < PREORDER(p))). Recall that for f = (p,q) in E_B, LOW1(F) =
PREORDER(q) and LOW2(f) = PREORDER(p). We can, in time linear in |E|
(see CHAPTER 1, FIGURE 1.24, LEXICOGRAPHIC BUCKET SORT) sort
these triples in lexicographic order. First in this list will be all edges incident
on the vertex v = x (the root of T) then all those edges f incident on the vertex
v with PREORDER(v) = 2, etc. For a fixed vertex p, the incident edges f of
T ∪ B will be sorted first according to LOW1(f). Next, given LOW1(f) =
LOW1(f′) for two edges f = (p,q), f′ = (p,q′) incident to p, f will come before
f′ if LOW2(f) ≥ PREORDER(p) but LOW2(f′) < PREORDER(p). Note that,
due to biconnectedness, if p ≠ x then LOW1(f) < PREORDER(p). Thus,
considering all edges of T ∪ B incident to a fixed vertex p and having the same
LOW1 value, all edges f = (p,q) with LOW2(f) ≥ PREORDER(p) come first
in the sorted list of edges. Within this class of edges (incident to p, same LOW1
value and LOW2 ≥ p) and within the complement of this class (incident to p,
same LOW1 value, LOW2 > p) the order is arbitrary depending on exactly how
the bucket sort was carried out.

Using the above linear order on edges, we convert the ordered, directed,
rooted graph T ∪ B into another such graph T̄ ∪ B̄ with only the order changed.
We call this graph, together with T and B, a *properly ordered decomposition*
of G. Thus to specify such a decomposition we need T = (V,E_T,x), an ordered,
directed, rooted, lineal, spanning tree; B = (V_B,E_B) the associated ordered
directed graph. The orders on T and B are inherited from the order on G and

the order on $\tilde{T} \cup \bar{B}$ is as just described above. The functions PREORDER and POSTORDER still refer to T, not the new ordered tree $\tilde{T}$.

7.8 CONVENTION.

By using the function PREORDER, we can replace each vertex $v \in V$ by the integer PREORDER(v). *Henceforth we assume that* v = PREORDER(v) (relative to T).

We now discuss an example. Consider the graph embedding of FIGURE 7.9. The vertices have been given the PREORDER labeling relative to the ordered spanning tree T. T and B are not shown explicitly in FIGURE 7.9 but can be reconstructed immediately from the PREORDER numbers and the given embedding. The graph $\tilde{T} \cup \bar{B}$ is shown in FIGURE 7.10. The edges E_B are dashed arrows, the edges E_T are solid arrows. The order on edges (of $\tilde{T} \cup \bar{B}$) incident on a vertex p is obtained by reading counterclockwise starting at 12 o'clock. The order on edges incident to p is the order in which they are encountered locally at p when read in this manner. We shall only be concerned with the order on edges e with TAIL(e) = p (the edges directed away from p). TABLE 7.11 shows the adjacency table for the ordered directed graph $\tilde{T} \cup \bar{B}$.

7.9 A GRAPH EMBEDDING WITH PREORDER LABELS.

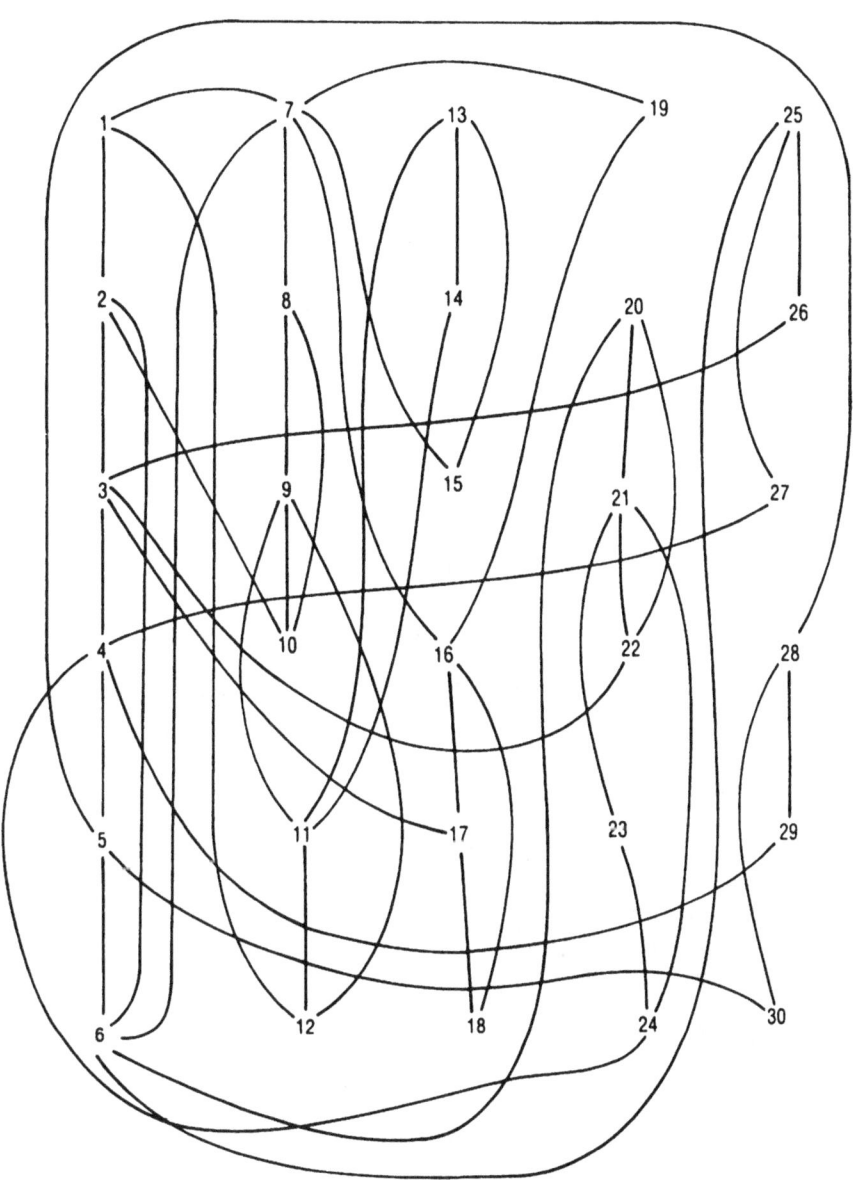

Figure 7.9

7.10 A PROPERLY ORDERED DECOMPOSITION $\tilde{T} \cup \tilde{B}$ OF THE GRAPH OF FIGURE 7.9.

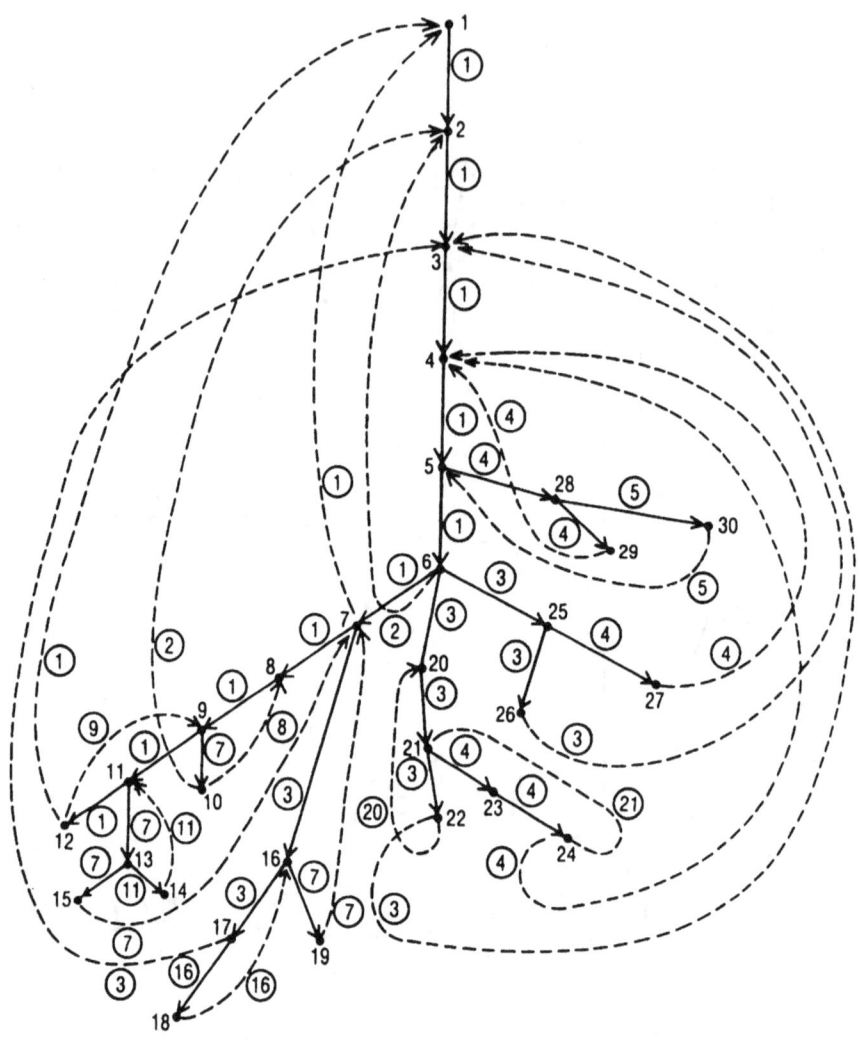

Circled edge labels are LOW1 values

Figure 7.10

7.11 ADJACENCY TABLE FOR T̄ ∪ B̃ OF FIGURE 7.10.

Table 2.1

1	2	16	17,19
2	3	17	3,18
3	4	18	16
4	5	19	7
5	6,28	20	21
6	7,2,20,25	21	22,23
7	1,8,16	22	3,20
8	9	23	24
9	11,10	24	4,21
10	2,8	25	26,27
11	12,13	26	3
12	1,9	27	4
13	15,14	28	29,30
14	11	29	4
15	7	30	5

7.12 REMARKS.

(1) The order of directed edges incident on the vertex 6 in FIGURE 7.10 is
(6,7),(6,2),(6,20),(6,25). Notice that LOW2(6,20) = 4 and LOW2(6,25)
= 4. Thus $\chi($LOW2(6,20) < PREORDER(6) = 6$) = 1$ and $\chi($LOW2(6,25)
< 6$) = 1$. By our previous remarks, the order of these two edges could
have been reversed in the ordering on T̄.

(2) At vertex 9 the order on edges of T̄ ∪ B̃ is (9,11),(9,10). Both edges are
in T̄. According to the definition of PREORDER, the order of these edges
is reversed in T. Thus the orders on T and T̄ are not the same.

Given a properly ordered decomposition T, B, T̄ ∪ B̃ of an ordered graph
G, we define a new tree PATR(G,T) called the *tree of paths* of G relative to T.
This tree will be needed later to give a global description of the basic recursive
structure that we shall use to construct the embeddings of a planar graph. Each
path will in fact define a cycle in G. Thus the "tree of paths" defines a "tree
of cycles" which is, in fact, a very special case of the tree of cycles of
DEFINITION 6.107 and FIGURE 6.108. The special nature of the tree of paths
is critical in the construction of an efficient planarity algorithm. To each edge
e = (a,b) in G we assign a path, PATH(e). If e ∈ E_B then PATH(e) ← (e). If
e ∉ E_B then PATH(e) will be a sequence of edges PATH(e) = $(e_1,e_2,. . .,e_k)$
defined by *procedure* 131 (if e = (a,b), define a = TAIL(e), b = HEAD(e)).

7.13 *procedure* PATH(e). [e = (TAIL(e), HEAD(e))]

begin

    if e $\in$ E$_B$ *then*

        PATH(e) $\leftarrow$ (e)

    else

        begin

            f $\leftarrow$ first edge of $\bar{T} \cup \bar{B}$ incident on and directed from HEAD(e); PATH(e) $\leftarrow$ (e,PATH(f)) [e added as first edge to PATH(f)];

        end

end

Thus to obtain PATH(e), e $\in$ T, one follows the sequence of first available edges (with respect to the ordering of $\bar{T} \cup \bar{B}$) until one encounters an edge in B. If PATH(e) = $(e_1, \ldots, e_k)$, with $e_i = (a_i, b_i)$, then notice that $b_k = $ LOW1(e) (as a consequence of the ordering on $\bar{T} \cup \bar{B}$) and LOW1(e) $< a_1$ if $a_1 \neq$ x (by biconnectedness of G). The reader should recall that we have assumed that for all v $\in$ V, v = PREORDER(v). If we addend to each PATH(e) the unique sequence of tree edges from b_k to a_1 we obtain a cycle that we call CYCLE(e). CYCLE(e) may be computed easily from the function FATHER discussed previously, once PATH(e) is known. The map PATH(e) $\rightarrow$ CYCLE(e) will convert the tree of paths (defined below) to a tree of cycles in the sense of DEFINITION 6.107.

7.14 DEFINITION.

We now define the directed ordered rooted "path tree," PATR(G,T). The vertices of PATR(G,T) will be a *subset* of the set {PATH(e): e $\in$ E(G)}. PATR(G,T) will be an ordered rooted tree. Since T is lineal and G is biconnected, there is only one edge d of T incident on the root x of T. The root of PATR(G,T) is PATH(d). In general, if PATH(e) is a vertex of PATR(G,T) then let PATH(e) = $(e_1, e_2, \ldots, e_k)$ where $e_i = (a_i, b_i)$. For each i = 1, \ldots, k, let $\bar{E}_{a_i}$ denote the ordered (in $\bar{T} \cup \bar{B}$) list of edges incident to a_i and let $\bar{E}[e_i] = \bar{E}_{a_i} - \{e_i\}$ be this list with e_i (the first entry) removed. Let $\mathscr{E}(e)$ denote the list of edges obtained by concatenating $\bar{E}[e_i]$, i $\geq$ 2, in reverse order: $\bar{E}[e_k]$, $\bar{E}[e_{k-1}], \ldots, \bar{E}[e_2]$. In particular, if PATH(e) = (e) then $\mathscr{E}(e)$ is empty. The ordered list of edges incident on PATH(e) in PATR(G,T) is {(PATH(e),PATH(f)): f $\in$ $\mathscr{E}(e)$} where the order is that included by the order on $\mathscr{E}(e)$. In this manner we recursively construct PATR(G,T). (It is easily verified that the directed graph constructed by this process is a tree.)

FIGURE 7.15 gives PATR(G,T) for the example of FIGURE 7.10. If PATH(e) = $(e_1, \ldots, e_k)$ with $e_i = (a_i, b_i)$, then the sequence $a_1, \ldots, a_k b_k$ specifies PATH(e).

In each vertex of FIGURE 7.15, this sequence is written in reverse order to facilitate the planar representation of the tree. (For visual clarity, we have shown edges between vertices connecting the starting value a_1 of each son to the occurrence of a_1 in the string representing its father. This is intended as an aid to reading the tree and is not a part of the definition of PATR(G,T).) Note that each backedge occurs in exactly one path so the number of vertices of PATR(G,T) is $|E_B| = |E| - |V| + 1$ (the *nullity* of G, n(G), DEFINITION 6.90).

7.15 PATR(G,T) FOR FIGURE 7.10.

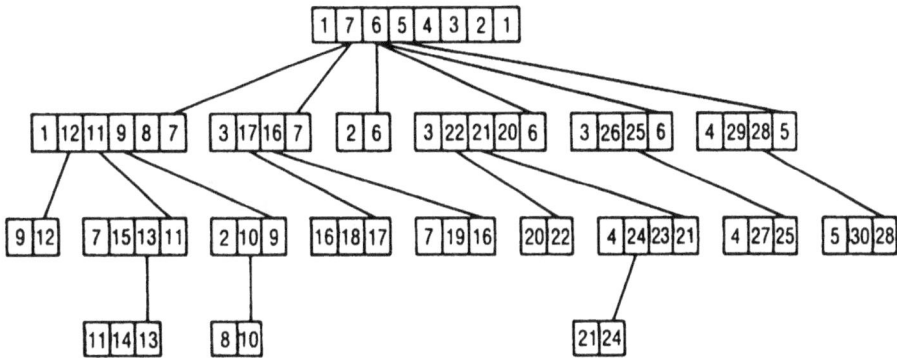

Note: PATH(e) $= (a_1,b_1) (a_2,b_2) \ldots (a_k,b_k)$ with $b_1 = a_2$, $b_2 = a_3, \ldots,$ is given by the sequence $\boxed{b_k|a_k|a_{k-1}|\ldots|a_1}$.

Figure 7.15

Consider PATR(G,T) (DEFINITION 7.14, FIGURE 7.15). $G = (V,E)$ is as above with PREORDER(v) $=$ v for all v $\in$ V. Observe that the set {PATH(e): PATH(e) a vertex of PATR(G,T)} defines a (set) partition of the set of edges E of G (the "blocks" of the partition are the sets of edges of the various paths, PATH(e)). The union of all subgraphs of $\tilde{T} \cup \tilde{B}$ of the form PATH(f) for PATH(f) a descendant of PATH(e) (including PATH(e)) in PATR(G,T) is a subgraph of $\tilde{T} \cup \tilde{B}$ which we call the *segment* of e in $\tilde{T} \cup \tilde{B}$. We denote this subgraph by SEG(e). The edges of SEG(e) contained in E_B will be called the *backedges* of SEG(e) and those contained in E_T will be called the *tree edges*. SEG(e) consists of all edges of T descendant to $e = (a,b)$ and all backedges incident on vertices of T descendant to b. SEG(e) is ordered as a subgraph of $\tilde{T} \cup \tilde{B}$. When we speak of planar embeddings of SEG(e) we do not require that they reflect the ordering of $\tilde{T} \cup \tilde{B}$. From the definition of PATR(G,T) it is apparent that if (s,t) is a backedge of SEG(e), and PATH(e$'$) is the father of PATH(e) in PATR(G,T), then t $\geq$ LOW1(e) $\geq$ LOW1(e$'$). Thus t is either a vertex of SEG(e) not on CYCLE(e$'$) or a vertex of CYCLE(e$'$) intersected with CYCLE(e). See

79

FIGURE 7.16. Note that SEG(e) is a bridge of CYCLE(e') in the sense of DEFINITION 6.100 and FIGURES 6.98 and 6.99.

7.16 THE BRIDGE SEG(e) FOR CYCLE(e').

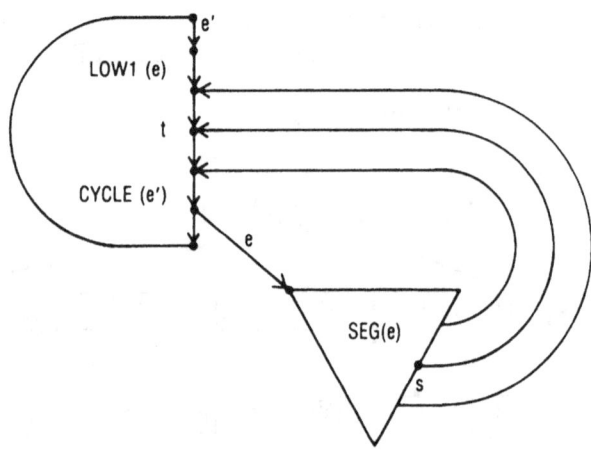

Figure 7.16

Let $e = (a,b)$ and consider all backedges (s,t) of SEG(e) with $LOW1(e) < t < a$. Call such backedges of SEG(e) *proper*. Observe from FIGURE 7.16 that if G is planar then the graph CYCLE(e) ∪ SEG(e) must be planar, and in any embedding all proper backedges must lie *inside* CYCLE(e) (as CYCLE(e') ∪ CYCLE(e) ∪ SEG(e) is planar). We call a planar embedding of CYCLE(e) ∪ SEG(e) *consistent* if all proper backedges of SEG(e) lie inside CYCLE(e). FIGURE 7.17 shows an example where CYCLE(e) ∪ SEG(e) is planar but has no consistent embedding. If $e = (x,b)$ is the (unique by biconnectedness) edge of $\tilde{T}$ incident on the root x of $\tilde{T}$ then SEG(e) has no proper backedges ($LOW1(e) = x$). Thus in this case $\tilde{T} \cup \tilde{B} = SEG(e)$ has a consistent embedding if and only if G is planar. Thus we have the following basic observation: G *is planar if and only if for every vertex PATH(e) of PATR(G,T) the graph* CYCLE(e) ∪ SEG(e) *has a consistent embedding.* The reader will note that CYCLE(e') ∪ CYCLE(e) ∪ SEG(e) is a $\mathscr{C}$-bicomponent of the cycle $\mathscr{C} = $ CYCLE(e') in the sense of DEFINITION 6.101. CYCLE(e) ∪ SEG(e) is the carrier of the bridge SEG(e) as defined in DEFINITION 6.102. If each vertex PATH(e) in PATR(G,T) is replaced by CYCLE(e), we have a cycle tree in the sense of DEFINITION 6.107.

80

7.17 PLANAR BUT NOT CONSISTENT.

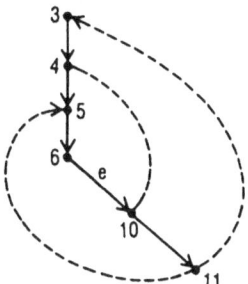

CYCLE(e) = (6,10,11,3,4,5,6)
Proper backedges: (10,4), (11,5)
e = (6,10)   LOW1(e) = 3.

Figure 7.17

We introduce some additional basic terminology. Given PATH(e), a vertex of PATR(G,T), let SEGLST(e) (called the *list of segments of* e) denote the set {SEG(f): PATH(f) a son of PATH(e) in PATR(G,T)}. Regard SEGLST(e) as linearly ordered by the order on PATR(G,T). Referring to FIGURE 7.15, let PATH(e) = PATH((7,8)) = $\boxed{1|12|11|9|8|7}$. Then SEGLST((7,8)) is (SEG((12,9)), SEG((11,13)), SEG((9,10))). These graphs may be read off directly from FIGURE 7.15. Let X, Y be segments in SEGLST(e). We say that X and Y are *directly linked* if, in any planar embedding of X ∪ Y ∪ CYCLE(e), X and Y must be on opposite sides of CYCLE(e). We define an undirected graph SEGGR(e) with vertex set SEGLST(e) and edge set {{X,Y}: X directly linked to Y}. SEGGR(e) is called the *segment graph* of e. In general, the number of edges of the segment graph may be quadratic in |V|. FIGURE 7.19 shows an example where the segment graph is the complete bipartite graph $K_{3,3}$. An obvious extension gives $K_{n,n}$ as a segment graph for |V| = 2n+3. The reader should note that the SEGGR(e) is a *subgraph* of the more general bridge graph of CYCLE(e) as defined in DEFINITION 6.110.

It is clear that if G is planar then every SEGGR(e) associated with PATR(G,T) must be bipartite or bichromatic. We "color" each vertex of SEGGR(e) with I if the segment defined by that vertex is inside CYCLE(e) in a given planar embedding or with O if it is outside.

Due to the above remarks about consistent embeddings, the converse is not quite true. Let us define a segment X in SEGLST(e), e = (a,b), to be *internal* if it has at least one backedge (s,t) with LOW1(e) < t < a. Let $\mathscr{I}$ denote the set of internal segments in SEGLST(e).

7.18   DEFINITION.

We say that SEGGR(e) is *$\mathcal{I}$-bichromatic* if there exists a 2-coloring of the vertices of SEGGR(e) where all vertices in $\mathcal{I}$ have the same color. Such a coloring will be called an *$\mathcal{I}$-bicoloring*.

7.19   COMPLETE BIPARTITE GRAPHS AS SEGMENT GRAPHS.

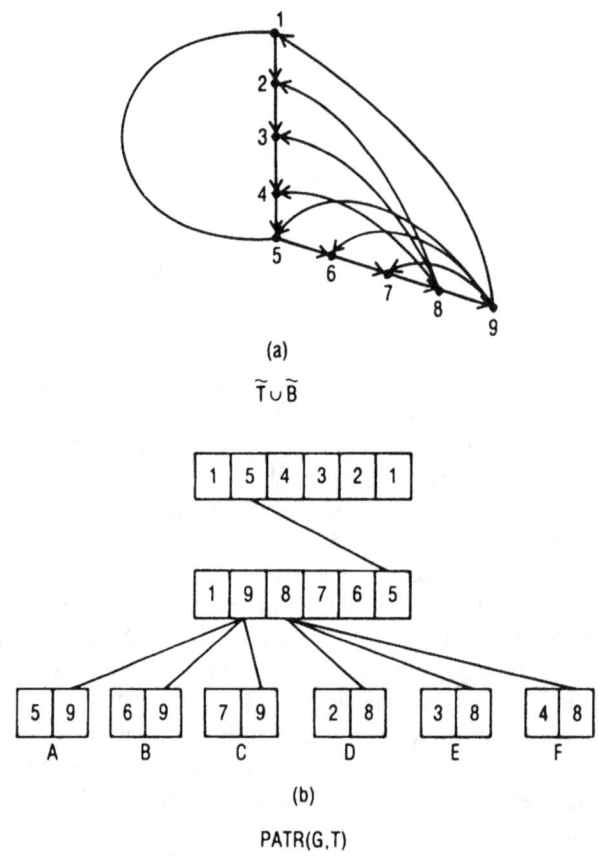

(a)

$\widetilde{T} \cup \widetilde{B}$

(b)

PATR(G,T)

Figure 7.19 (cont.)

82

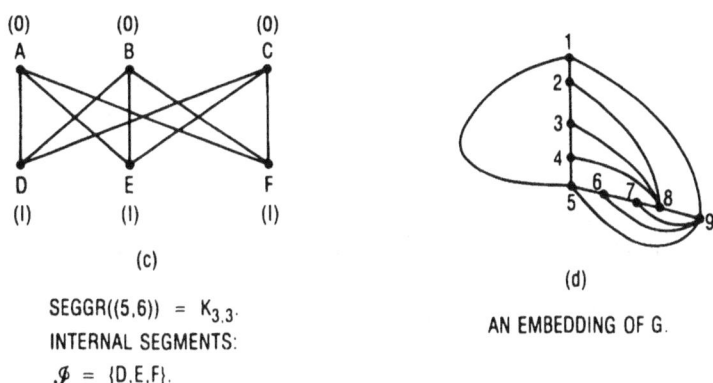

(c)

SEGGR((5,6)) = $K_{3,3}$.
INTERNAL SEGMENTS:
$\mathscr{I}$ = {D,E,F}.

(d)

AN EMBEDDING OF G.

Figure 7.19 (continued)

These observations are summarized in THEOREM 7.20.

7.20 THEOREM.

G is planar if and only if every SEGGR(e) associated with PATR(G,T) is $\mathscr{I}$-bichromatic where $\mathscr{I}$ is the set of internal segments of SEGLST(e).

For example, consider FIGURE 7.19. If e = (5,6) and $\tilde{T} \cup \tilde{B}$ is as shown in FIGURE 7.19(a), then there are six segments in SEGLST(e). Each segment is a simple backedge as indicated in FIGURE 7.19(b). We label these segments A through F. D, E, and F are the internal segments so $\mathscr{I}$ = {D,E,F}. We make the convention that when we color a segment graph, SEGGR(e), we use the two "colors" I and O. Elements of $\mathscr{I}$ will always be colored with I. FIGURE 7.19(c) shows an $\mathscr{I}$-coloring of SEGGR(e) for this example, so SEGGR(e) is $\mathscr{I}$-bichromatic. SEGGR(1,2) consists of one vertex, SEG(5,6), and no edges. $\mathscr{I}$ = ϕ for this case. Thus all segment graphs are $\mathscr{I}$-bichromatic. An embedding of G (defined by $\tilde{T} \cup \tilde{B}$) is shown in FIGURE 7.19(d).

To specify a planar embedding for $\tilde{T} \cup \tilde{B}$ (order in $\tilde{T} \cup \tilde{B}$ ignored in the embedding) it suffices to construct SEGGR(e) for each vertex of PATR(G,T) and $\mathscr{I}$-bicolor it (let us say with I and O, vertices in $\mathscr{I}$ colored I). Assuming recursively, that for all vertices (i.e., all SEG(e) $\in$ SEGLST(e)) of SEGGR(e) we have specified a consistent embedding, then the $\mathscr{I}$-bicolored SEGGR(e) specifies how to orient these embeddings relative to CYCLE(e) to construct a consistent embedding of SEG(e) relative to CYCLE(e). If at any stage in constructing the graphs SEGGR(e) we find one that is not $\mathscr{I}$-bichromatic then G is nonplanar. As remarked above, we cannot in general compute the SEGGR(e) in linear time in |V|. Nevertheless, this process provides a quite effective means for constructing planar embeddings of moderately complex graphs by "pencil

83

and paper" computation. One can show that a *spanning forest* for SEGGR(e) can be computed in linear time in vertices of G and this leads to a linear time planarity algorithm along these lines.

Recalling FIGURE 7.16, one can avoid the notion of "$\mathcal{I}$-bichromatic" in THEOREM 7.20 if one adds CYCLE(e')\CYCLE(e) to the SEGLST(e). Thus, one treats CYCLE(e')\CYCLE(e) as just another segment relative to CYCLE(e). With this slightly extended notion of SEGGR(e) the term "$\mathcal{I}$-bichromatic" of THEOREM 7.20 can be replaced with "bichromatic." We prefer the notion of $\mathcal{I}$-bichromatic in this case as it is more descriptive of the actual computations involved. (CYCLE(e')\CYCLE(e) denotes edges of CYCLE(e') not in CYCLE(e)).

We now illustrate by example the use of THEOREM 7.20 in testing a graph G for planarity and constructing a planar embedding if one exists.

7.21 EXAMPLE OF TESTING FOR PLANARITY AND CONSTRUCTING AN EMBEDDING.

Consider $\bar{T} \cup \bar{B}$ of FIGURE 7.10 and PATR(G,T) shown in FIGURE 7.15. Inspect the vertices of PATR(G,T) in the depth-first sequence (i.e., generate DFV(PATR(G,T)). FIGURE 7.22 shows PATR(G,T) with the vertices replaced by their PREORDER numbers. We use these numbers to refer to the vertices of PATR(G,T).

7.22 PREORDER NUMBERS FOR PATR(G,T) OF FIGURE 7.15.

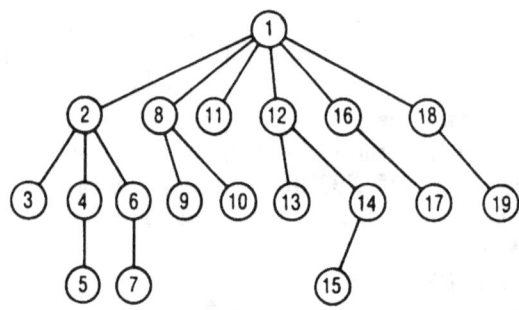

Figure 7.22

The vertices of FIGURE 7.22 listed in POSTORDER are 3, 5, 4, 7, 6, 2, 9, 10, 8, 11, 13, 15, 14, 12, 17, 16, 19, 18, 1. For each such integer k, let SEG(k), CYCLE(k), PATH(k), SEGGR(k), etc., denote the segment, cycle, path, segment graph, etc., associated with the vertex PATH(e) of PATR(G,T) having preorder number k. We attempt to construct an $\mathcal{I}$-coloration of SEGGR(k) and a consistent embedding of SEG(k) for the values of k as they occur in POSTORDER:

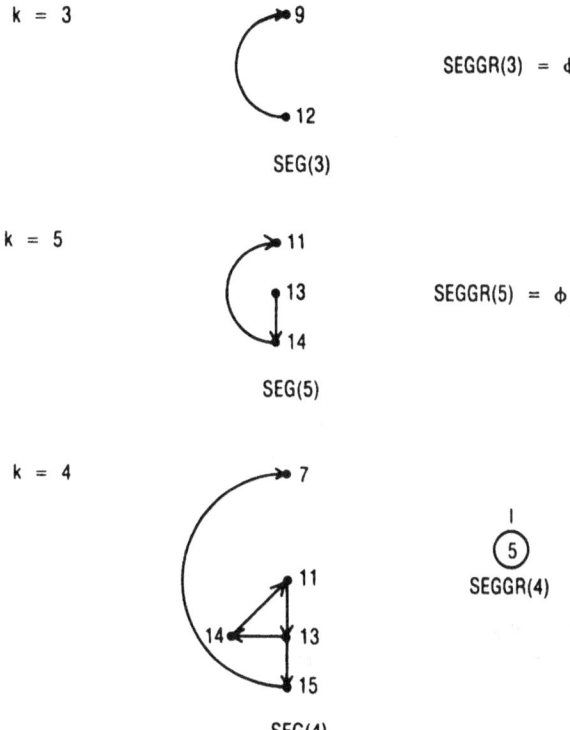

k = 3   9   SEGGR(3) = φ

12

SEG(3)

k = 5   11   SEGGR(5) = φ

13

14

SEG(5)

k = 4   7

11

14   13

15

SEG(4)

⑤
SEGGR(4)

Remark. "⑤" refers to "SEG(5)" in SEGGR(4). The manner of embedding SEG(5) relative to CYCLE(4) is left to trial and error at this stage, $\mathcal{I} = \phi$ here.

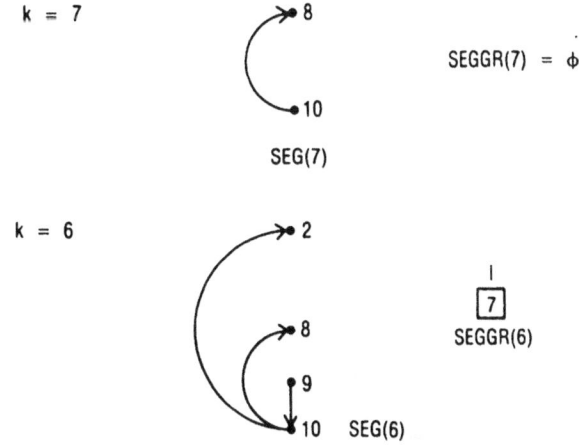

k = 7   8   SEGGR(7) = φ

10

SEG(7)

k = 6   2

7
SEGGR(6)

8

9

10   SEG(6)

Remark. SEG(7) is internal. This is indicated by □ in SEGGR(6).

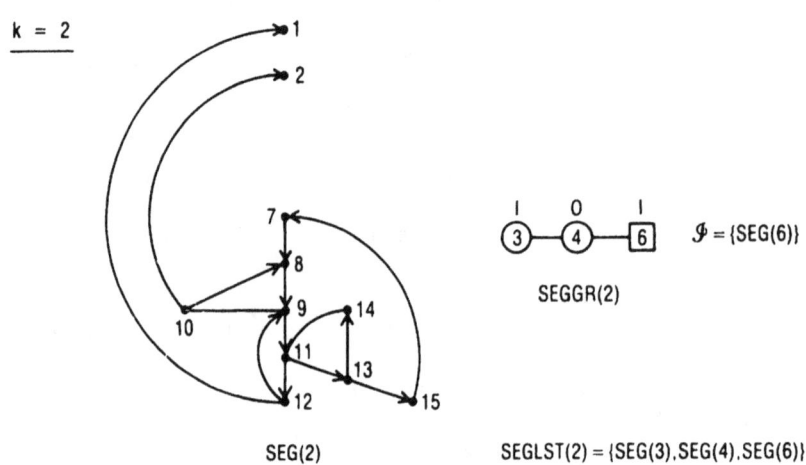

SEGGR(2)

SEG(2)　　　　　　SEGLST(2) = {SEG(3),SEG(4),SEG(6)}

Remark. We have already (because of postorder computation) given consistent embeddings of each element of SEGLST(2). SEG(3) and SEG(4) are *directly linked* (by inspection), as are SEG(4) and SEG(6).

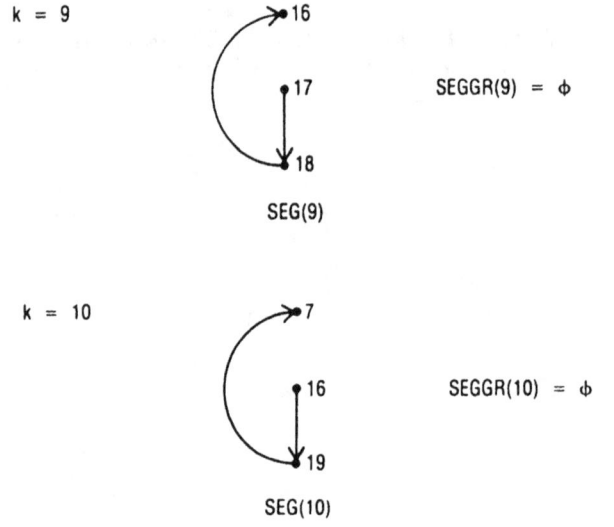

86

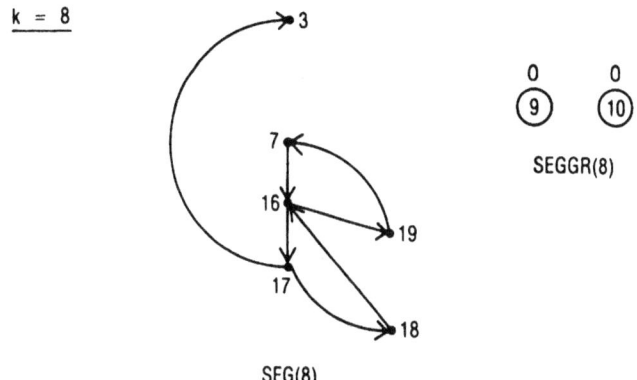

k = 8

SEGGR(8)

SEG(8)

Remark.   The two vertices can be colored I or O independently, so there are three other possible $\mathcal{I}$-colorations of SEGGR(8) ($\mathcal{I} = \phi$).

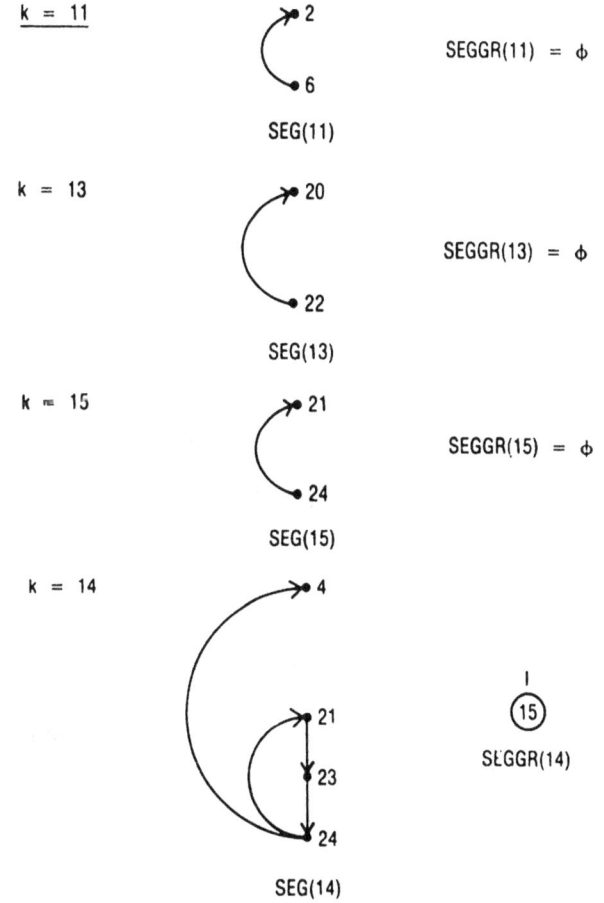

k = 11

SEG(11)

SEGGR(11) = ϕ

k = 13

SEG(13)

SEGGR(13) = ϕ

k = 15

SEG(15)

SEGGR(15) = ϕ

k = 14

SEG(14)

SŁGGR(14)

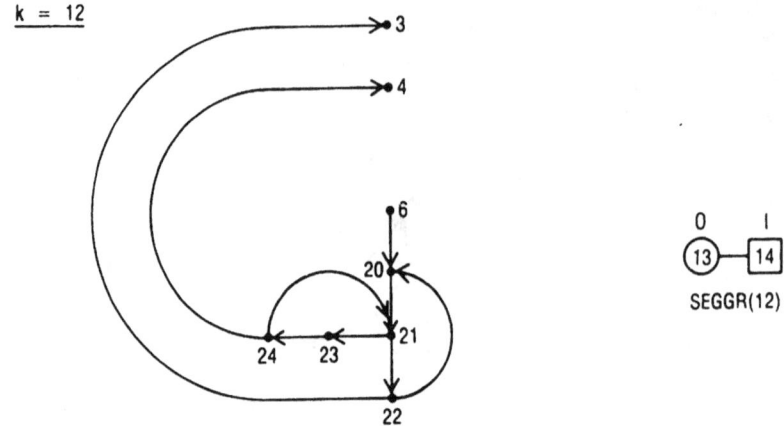

Remark. $\mathscr{I} = \{\text{SEG}(14)\}$. The $\mathscr{I}$-coloration of SEGGR(12) is unique as SEG(13) and SEG(14) are directly linked (determined again by inspection).

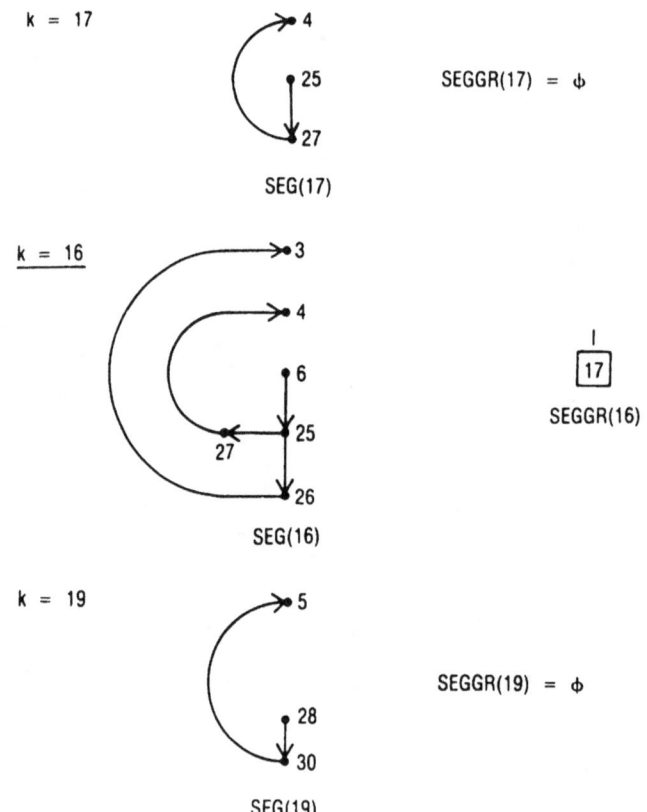

88

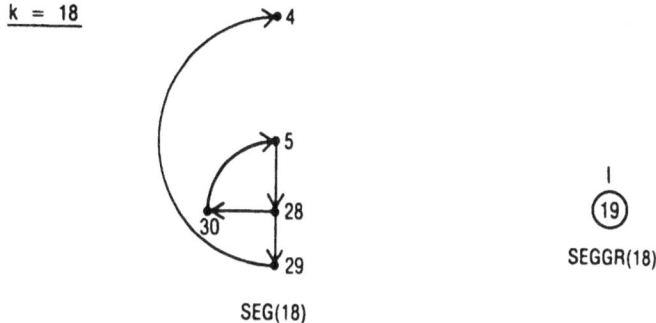

SEG(18)

SEGGR(18)

k = 1 *Remark.*   Sons of 1 in PATR(G,T): 2, 8, 11, 12, 16, 18 are underlinked above. Only the *proper* backedges of these segments (see FIGURE 7.17) and those backedges (s,t) with t = LOW1(e) affect the construction of SEGGR(1) and the subsequent embedding. This is shown in FIGURE 7.23.

7.23 SEGGR(1) AND ITS EMBEDDING.

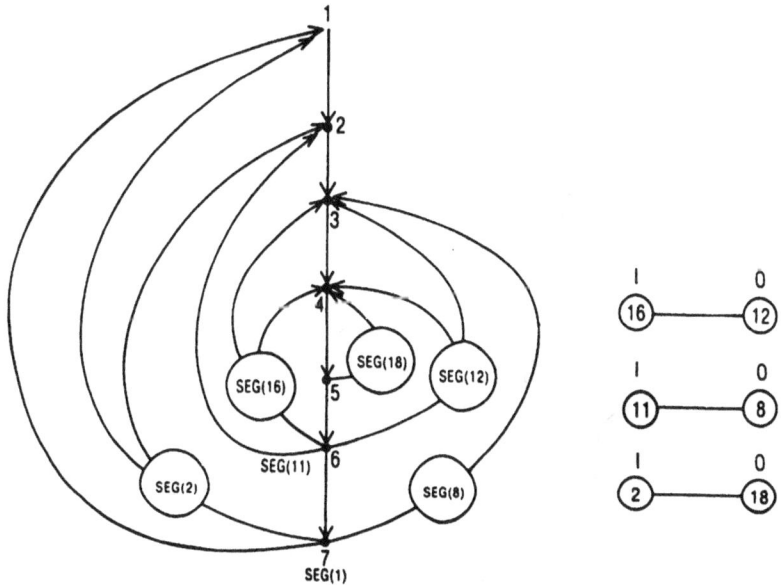

Figure 7.23

By filling in the detailed structure of the segments of the embedding shown in FIGURE 7.23 we obtain, in FIGURE 7.24, a planar embedding of the graph of FIGURE 7.9.

89

7.24 DETAILED RECONSTRUCTION OF FIGURE 7.23 AND PLANAR EMBEDDING OF FIGURE 7.9.

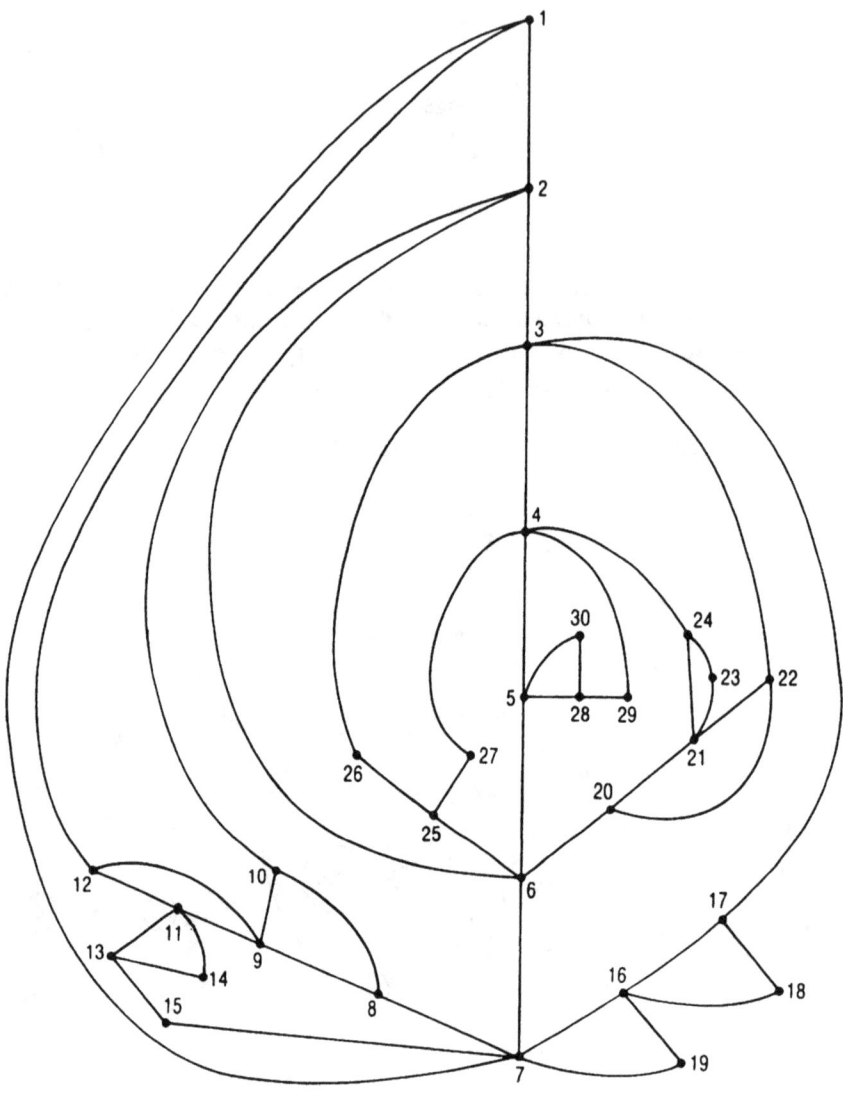

Figure 7.24

The reader should observe that the bicolorations of the segment graphs specify a labeling (with I,O) of PATR(G,T) as shown in FIGURE 7.25.

7.25 I,O LABELING OF PATR(G,T) OF FIGURES 7.15 AND 7.22.

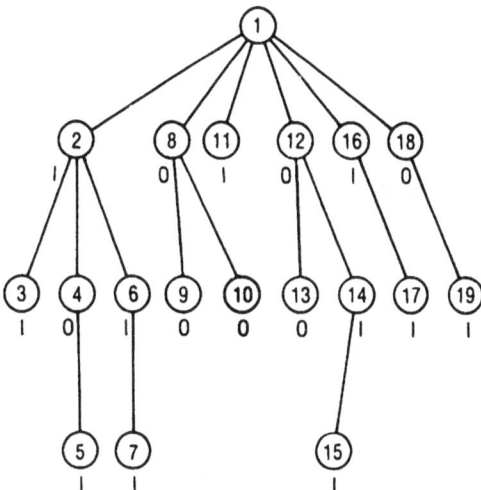

Figure 7.25

By traversing this labeled tree and, using the labels I = Inside, O = Outside, to embed the vertices of PATR(G) (as paths in G) in the plane as they are encountered in PREORDER (refer to FIGURE 7.15) we again obtain FIGURE 7.24. Thus FIGURE 7.25 may be regarded as a specification of an embedding.

We now consider two basic lemmas that aid in carrying out the type of computation illustrated by EXAMPLE 7.21. With the aid of these lemmas, the reader will find it not too difficult to set up the data structures required to test for planarity in linear time in the number of vertices. Recall that it is only necessary to consider the case where $|E| \leq 3|V| - 6$ (THEOREM 6.115). As defined above, if $SEG(f_1)$ and $SEG(f_2)$ are in $SEGLST(e)$ then we say $SEG(f_1)$ and $SEG(f_2)$ are *directly linked* if $SEG(f_1)$ and $SEG(f_2)$ must be embedded on opposite sides of $CYCLE(e)$ in any planar embedding of $CYCLE(e) \cup SEG(f_1) \cup SEG(f_2)$. We write $SEG(f_1)$ dl $SEG(f_2)$. Recall also that if $e = (a,b)$ is a directed edge then $TAIL(e) = a$, $HEAD(e) = b$.

7.26 LEMMA.

Let $SEG(f_1) < SEG(f_2)$ be two segments in $SEGLST(e)$. Then $SEG(f_1)$ is directly linked to $SEG(f_2)$ if and only if there is a backedge (s,t) of $SEG(f_1)$ with $LOW1(f_2) < t < TAIL(f_2)$.

LEMMA 7.26 is easily proved by considering cases. We note only the case where $TAIL(f_1) = TAIL(f_2)$ and $LOW1(f_1) = LOW1(f_2)$. In this case $LOW1(f_1)$

91

$< t < TAIL(f_1)$ so (s,t) is a *proper* backedge of $SEG(f_1)$ (as defined in connection with FIGURES 7.16 and 7.17). Recalling the definition of the order on SEGLST(e) as derived from the order on PATR(G,T) and the sorting of the triple

$$(PREORDER(p), LOW1(f), X(LOW2(f) < PREORDER(p))$$

$$\text{where } f = (p,q),$$

we see that $SEG(f_2)$ must also have a proper backedge. That is,

$$X(LOW2(f_2) < PREORDER(TAIL(f_2))) = 1.$$

This is the only place where the third component of the triple plays a role in what we have done. The fact that $SEG(f_1)$ dl $SEG(f_2)$ in this case is an immediate consequence of the existence of this proper backedge of $SEG(f_2)$. LEMMA 7.26 is a result of Hopcroft and Tarjan (see the references at the end of PART II).

7.27 EXERCISE.

(1) Complete the proof of LEMMA 7.26.
(2) What data structures would you use to implement the test of LEMMA 7.26 in the computation of EXAMPLE 7.21?

Lemma 7.26 provides the basis for efficiently constructing segment graphs or spanning forests for segment graphs insofar as deciding when a pair of segments forms an edge of the segment graph (i.e., when they are directly linked). The segments of SEGLST(e) are added one after another to SEGGR(e), checking at each stage to see if SEGGR(e) is still bipartite. This latter test is aided by the following ideas.

Consider FIGURE 7.29. CYCLE(e) = (1,2,. . .,20,1) for the edge e = (1,2) is shown. The segments of SEGLST(e) are labeled A,B,. . .,J in the order on SEGLST(e). The edges internal to these segments are not shown, but all backedges that go from a segment to CYCLE(e) are shown. Consider a particular segment, say H, and let SEGLST(e,H) denote all segments less than or equal to H in the ordering on SEGLST(e). Thus, SEGLST(e,H) = A,B,. . .,H in that order. This set of segments defines a subgraph of SEGGR(e) which we call SEGGR(e,H). The edges of this subgraph are defined by the ''directly linked'' relation just as in SEGGR(e). The components $\mathcal{H}_1$, $\mathcal{H}_2$, and $\mathcal{H}_3$ are shown in FIGURE 7.29. For any such component, let HEAD($\mathcal{H}$) denote the ordered sequence of vertices of CYCLE(e) of the form HEAD(f) where f is a backedge from some segment in $\mathcal{H}$ to CYCLE(e). For example, HEAD($\mathcal{H}_1$) = (1,2,3,17). In the same manner, let HEAD($\mathcal{H}$,H) denote those elements of HEAD($\mathcal{H}$) that are strictly less than the largest vertex of H on CYCLE(e) (we call this vertex TAIL(H)). In the example, TAIL(H) = 10 and HEAD($\mathcal{H}_1$,H) = (1,2,3). We say that a component $\mathcal{H}$ of SEGGR(e,H) is *relevant* if HEAD($\mathcal{H}$,H) is nonempty, otherwise $\mathcal{H}$ is called *irrelevant*. In the example, $\mathcal{H}_1$ and $\mathcal{H}_3$ are relevant to H but $\mathcal{H}_2$ is irrelevant. The basic observation that we need concerns the sets HEAD($\mathcal{H}$,H) for relevant components. In the example, HEAD($\mathcal{H}_1$,H) = (1,2,3) and HEAD($\mathcal{H}_3$,H) = (3,4,5,7,8). Note that the maximum value (last value) of the first sequence is less than or equal to the minimum value of the second sequence (in this example we have equality). In other words, the ordered sequences HEAD($\mathcal{H}$,H), as $\mathcal{H}$ ranges over all relevant components, don't overlap except possibly at their endpoints. This idea is stated in LEMMA 7.28. The order on vertices of CYCLE(e) is the natural order for integers as we are using the preorder numbers to label vertices. We order the components $\mathcal{H}_1,\mathcal{H}_2,\ldots,\mathcal{H}_p$ by their minimal segments (thus, $\mathcal{H}_1 < \mathcal{H}_2 < \mathcal{H}_3$ because A < D < F in SEGLST(e) in FIGURE 7.29). If the maximum of HEAD($\mathcal{H}_i$,H) is less than or equal to the minimum of HEAD($\mathcal{H}_{i+1}$,H) for i = 1,2,$\ldots$,p$-$1 we say that the sequence HEAD($\mathcal{H}_1$,H),$\ldots$,HEAD($\mathcal{H}_p$,H) is a *max-min chain* in CYCLE(e).

7.28 LEMMA.

Let H be any segment of SEGLST(e) and let $\mathcal{H}_1,\ldots,\mathcal{H}_p$ be the relevant components of SEGGR(e,H) ordered by their minimal segments. Then the sequence HEAD($\mathcal{H}_1$,H),$\ldots$,HEAD($\mathcal{H}_p$,H) is a max-min chain in CYCLE(e).

The proof of LEMMA 7.28 is left as an exercise for the reader, (EXERCISE 7.31(1)). The reader should note that LEMMA 7.28 does not involve any assumptions of planarity. The notion of a planar embedding enters only through the idea of ''directly linked'' in connection with defining the components $\mathcal{H}_i$. Referring again to FIGURE 7.29, note that SEGGR(e,H) is $\mathcal{I}$-bipartite ($\mathcal{I}$ is

7.29 MAX-MIN CHAIN.

$(\text{HEAD}(\mathcal{H}_1,H),\ \text{HEAD}(\mathcal{H}_3,H)) = ((1,2,3),(3,4,5,7,8))$.

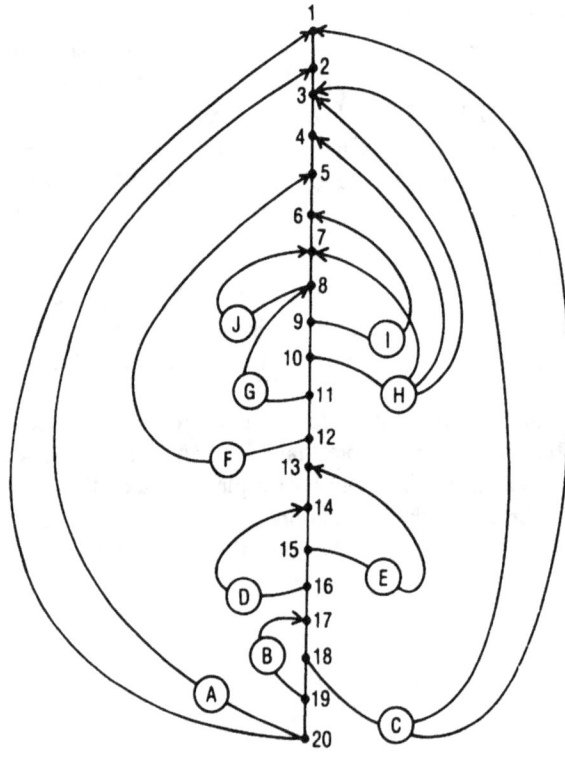

SEGLST$((1,2),H) = (A,B,\ldots,H)$
SEGGR$((1,2),H)$:

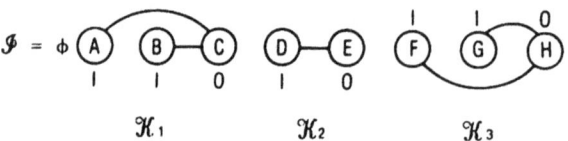

Figure 7.29

94

empty in this case) and that the embedding shown of SEGLST(e,H) together with CYCLE(e) is planar. This embedding gives the I,O labeling of the vertices of SEGGR(e,H) shown in FIGURE 7.29. Given this situation, we can split each sequence HEAD($\mathcal{H}$,H) into two subsequences, INHEAD($\mathcal{H}$,H) and OUT-HEAD($\mathcal{H}$,H). The sequence INHEAD($\mathcal{H}$,H) consists of all entries of HEAD($\mathcal{H}$,H) that are equal to HEAD(f) for some backedge f belonging to a segment of $\mathcal{H}$ that is labeled I (i.e., is embedded inside CYCLE(e)). OUTHEAD($\mathcal{H}$,H) is similarly defined. In FIGURE 7.29, INHEAD($\mathcal{H}_3$,H) = (5,8) and OUT-HEAD($\mathcal{H}_3$,H) = (3,4,7). For segment H, we define SPAN(H) to be the sequence of vertices of CYCLE(e) from LOW1(H) to TAIL(H) (see EXERCISE 6.109(3)). Thus, in FIGURE 7.29, SPAN(H) = (3,4,5,6,7,8,9,10) and SPAN(I) = (6,7,8,9). The *open* or *interior* span of H is SPAN(H) with its first and last entries removed. We denote this sequence by OSPAN(H). This sequence may be empty. In FIGURE 7.29, OSPAN(H) = (4,5,6,7,8,9), OSPAN(I) = (7,8). The critical observation here (FIGURE 7.29 again) is that both INHEAD($\mathcal{H}_3$,H) and OUT-HEAD($\mathcal{H}_3$,H) have entries in OSPAN(I). This means that when I is added to SEGGR(e,H) to form SEGGR(e,I) the new graph will not be $\mathcal{I}$-bipartite. In other words, SEGLST(e,H) together with CYCLE(e) is planar but SEGLST(e,I) together with CYCLE(e) is not planar. We state this observation as LEMMA 7.30. The reader should recall the relationship between CYCLE(e), CYCLE (e') and SEG(e) of FIGURE 7.16 and the notion of an internal segment defined in connection with DEFINITION 7.18.

7.30 LEMMA.

Suppose that H and I are consecutive elements of SEGLST(e) and that SEGLST(e,H) together with CYCLE(e) is planar. Suppose that an $\mathcal{I}$-bicoloration of SEGGR(e,H) has been specified (or a planar embedding of SEGLST(e,H) together with CY-CLE(e) has been specified). If there is a component $\mathcal{H}$ of SEGGR(e,H) such that both INHEAD($\mathcal{H}$,H) and OUTHEAD($\mathcal{H}$,H) have points in common with OSPAN(I), then SEGGR(e,I) is not $\mathcal{I}$-bichromatic (SEGLST(e,I) together with CYCLE(e) and hence the original graph G is not planar). If there is a component $\mathcal{H}$ of SEGGR(e,H) with a vertex A that is an internal segment and if I is an internal segment, then INHEAD ($\mathcal{H}$,H) $\cap$ OSPAN(I) $\neq$ 0 implies that SEGGR(e,I) is not $\mathcal{I}$-bichromatic (SEGLST(e,I) $\cup$ CYCLE(e) $\cup$ CYCLE(e')) and hence G is not planar).

The proof of LEMMA 7.30 is essentially a restatement of LEMMA 7.26. A few sketches of cases will convince the reader of its validity.

The reader who has thought carefully about the basic data structures presented in this chapter and in PART I is now in a good position to develop any one of a number of variations of a linear time planarity algorithm.

7.31 EXERCISE.

(1) Prove LEMMA 7.28. (*Hint:* One way is by induction on the linearly ordered set SEGLST(e). Suppose X and Y are consecutive in SEGLST(e) and that the lemma is true for X. Add Y to SEGGR(e,X) to get SEGGR(e,Y) and see what happens to the max-min chain.)

(2) Using the ideas of LEMMAS 7.26, 7.28, and 7.30 try to describe the data structures for a linear time planarity algorithm. Try and convince at least two friends that your algorithm works in linear time. (*Hint:* A major hurdle will be the soporific effect on the friends. Brew a pot of strong coffee!)

(3) Extend the previous exercise to compute a spanning forest SEGFO(e,H) for each SEGGR(e,H). Specify data structures and discuss complexity. In the discussion of complexity, distinguish between the planar and nonplanar cases.

The reader interested in pursuing the topic of planarity algorithms further should begin with the seminal work of Hopcraft and Tarjan cited in the references (*Efficient Planarity Testing*, J. Assoc. Comp. March 21 (1974), 549–568).

Index